In vitro techniques in research

In Vitro techniques in research

Recent advances

Edited by
J. W. PAYNE

Open University Press
in association with the Humane Research Trust

Open University Press
12 Cofferidge Close
Stony Stratford
Milton Keynes MK11 1BY

and

242 Cherry Street
Philadelphia, PA 19106, USA

First Published 1989

British Library Cataloguing in Publication Data
Payne, J.W. (John Weston)
In vitro techniques in research.
1. Organisms. Cells. Differentiation.
Laboratory techniques : In vitro methods
I. Title
574.87′612′028

ISBN 0-335-15885-4

Typeset by Vision Typesetting, Manchester
Printed in Great Britain by the Alden Press, Oxford

Contents

Foreword

The papers in this book formed the greatest part of The Humane Research Trust Conference entitled *Recent Advances in the Use of in vitro Techniques*. As Chairman of the Trust it was my honour to welcome distinguished delegates and speakers and to thank the Dean of the Hunterian Institute, Professor Graham P. Lewis, for enabling the meeting to take place in the august surroundings of The Royal College of Surgeons.

In sponsoring the Conference, the Trust was forwarding its purpose of encouraging innovative scientific medical research in which the use of the animal as a model for man is replaced by other techniques. Over a period of years assistance has been given to many remarkable programmes, some of which might never have started without this help. Frequently, Humane Research Trust support is given as seeding money for work which had, after the initial proving, gone on to achieve further support which otherwise, might not have been forthcoming.

It was a particular pleasure to welcome so many delegates from industry at the Conference and I am delighted that this book will bring the excellent papers to a wider audience.

On behalf of the Trustees I should also like to thank Professor Paul Turner of St Bartholomew's Hospital for his guidance.

Finally, and importantly, thanks to Professor John W. Payne of the University College of North Wales for generously accepting the task of editing the highly technical and detailed papers. To him and all the contributors, many thanks are owed.

R. MacAlastair Brown, *Chairman*

The Humane Research Trust
Brook House
Bramhall Cheshire SK7 2DN

Editor's preface

This book presents edited versions of most of the talks given at the conference entitled *Recent Advances in the Use of in vitro Techniques*, sponsored by The Humane Research Trust and held at the Royal College of Surgeons, London on 17 September 1987. In addition, there are several papers from authors who were unable to contribute to the conference. Unfortunately, it has not been possible to publish a record of the very many posters presented at the meeting.

Many of the authors have at some stage in their investigations received sponsorship from The Humane Research Trust. The contributions, therefore, reflect the range of scientific and medical research supported by the Trust in its endeavours to promote strategies aimed at finding alternatives to the use of animals in scientific research. However, the conference and now this book do more than merely document, for they testify to the effectiveness of the Trust in targeting its resources into fruitful areas. The papers describe novel ideas and innovative techniques that have contributed to significant developments in various areas; many of the authors are leading experts in their fields. Yet the Trust most often committed its funds at an early stage of their work, thereby playing a crucial pump-priming role that in many cases was subsequently endorsed by Research Council funding. It is to be hoped that this successful synergism between the scientific community and the Trust will continue and develop in the future.

All but the last of the papers deal primarily with eukaryotic cell systems, but they cover several different topics. In Chapter 1, Pearson and Chijioke briefly review a range of isolated human cells that has been used in *in vitro* studies as substitutes for animal tests. In the first of several papers using particular cell types, Lewis (Chapter 2) describes results obtained with cultured human mononuclear cells on the effects and mechanisms of action of anti-inflammatory and anti-rheumatic drugs. Bayliss and colleagues (Chapter 3) discuss the current and potential uses of cell and organ cultures of human cartilage material, which maintain the zonal activity and chemical composition that are present *in vivo*. J. N. Payne describes in Chapter 4 how the uptake and transport of exogenous materials by nerve cells can be studied in tissue culture. The studies are aimed at characterizing the retrograde axonal transport, i.e. the movement along the axon back to the neurone cell body, of, for example, trophic agents implicated in developmental processes. Carey and his colleagues (Chapter 5) describe their studies with cultured glial cells from neonatal rodent brain, aimed at characterizing the development of oligodendrocytes and their role in the process of myelination in the central nervous system. Polak and Bloom (Chapter 6) point out the advantages of using tumour cell lines, in which one can have large numbers of identical cells, in contrast to the inherent variability between animals, to study the intracellular events which take place during peptide hormone synthesis and release by neuroendocrine (tumour) cells. These studies highlight the application of powerful new techniques such as *in situ* hybridohistochemistry. Bach in Chapter 7 discusses application of varied microscopical techniques allowing just a few cells to be sufficient for certain studies in culture; thus, renal glomerular epithelial cells, proximal tubular cells and renal medullary interstitial cells have been used to study the molecular basis of targeting against specific cell types seen with certain chemicals *in vivo*. In a slight change of emphasis, Rhodes and colleagues (Chapter 8) review techniques for the assessment of acute toxicity, ranging from studies *in vivo*, through cell culture to tissue and organ culture. Here, *in vitro* techniques are advocated as pre-screening devices to assess the presence or absence of adverse effects of chemicals, which allows informed judgement on the need or desirability to proceed to the live animal stage. Garner (Chapter 9) continues this theme by considering the application of various subcellular, biochemical techniques as a means of assessing the exposure of humans to carcinogens; he also discusses the relevance of animal carcinogenicity experiments to human exposure studies. Bacteria are successfully used in many *in vitro* studies, e.g. the Ames test, and in Chapter 10, J. W. Payne describes the production of new strains of *Escherichia coli* for use in novel, enzyme-linked assays for nutritionally-available amino acids in foods; these methods offer preferred alternatives to current animal and chemical procedures.

Contributors

C. W. Archer Professorial Research Unit, Institute of Orthopaedics, Stanmore, Middlesex HA7 4LP

P. H. Bach The Nephrotoxicity Research Group, Robens Institute of Environmental and Industrial Health and Safety, University of Surrey, Guildford, Surrey GU2 5XH

M. T. Bayliss Division of Biochemistry, Kennedy Institute of Rheumatology, 6 Bute Gardens, Hammersmith, London W6 7DW

S. R. Bloom Departments of Histochemistry and Medicine, Royal Postgraduate Medical School, Hammersmith Hospital, Du Cane Road, London W12 0HS

E. M. Carey Department of Biochemistry, University of Sheffield, Sheffield S10 2TN

P. C. Chijioke Department of Clinical Pharmacology, St Bartholomew's Hospital Medical College, London EC1A 7BE

R. C. Garner Cancer Research Unit, University of York, Heslington, York YO1 5DD

N. Herschkowitz Department of Paediatrics, University of Berne, Switzerland

B. Johnstone Division of Biochemistry, Kennedy Institute of Rheumatology, 6 Bute Gardens, Hammersmith, London W6 7DW

G. P. Lewis Department of Pharmacology, Hunterian Institute, Royal College of Surgeons, Lincolns Inn Fields, London WC2A 3PN

J. McDowell Professorial Research Unit, Institute of Orthopaedics, Stanmore, Middlesex HA7 4LP

G. J. A. Oliver Imperial Chemical Industries plc, Central Toxicology Laboratory, Alderley Park, Macclesfield, Cheshire SK10 4TG

J. N. Payne Department of Anatomy and Cell Biology, Sheffield University, Sheffield S10 2TN

J. W. Payne Department of Biochemistry, School of Biological Sciences, University of Wales, Bangor, Gwynedd LL59 5NP

R. M. Pearson Department of Clinical Pharmacology, St Bartholomew's Hospital Medical College, London EC1A 7BE

M. A. Pemberton Imperial Chemical Industries plc, Central Toxicology Laboratory, Alderley Park, Macclesfield, Cheshire SK10 4TG

J. M. Polak Departments of Histochemistry and Medicine, Royal Postgraduate Medical School, Hammersmith Hospital, Du Cane Road, London W12 0HS

R. Reynolds Department of Biochemistry, Imperial College, London SW7 2BB

C. Rhodes Imperial Chemical Industries plc, Central Toxicology Laboratory, Alderley Park, Macclesfield, Cheshire SK10 4TG

R. C. Scott Imperial Chemical Industries plc, Central Toxicology Laboratory, Alderley Park, Macclesfield, Cheshire SK10 4TG

M. Vojvodic Department of Biochemistry, University of Sheffield, Sheffield S10 2TN

CHAPTER 1

In vitro studies with isolated human cells

R. M. Pearson and P. C. Chijioke

Current objections to the relevance and ethics of using large numbers of live experimental animals or human subjects to examine the efficacy and toxicity of drugs and chemicals has concentrated attention on the possible use of isolated human cells for these purposes. The development of tests with isolated cell systems *in vitro* in order to reduce or possibly eliminate the need for living animals depends on a series of assumptions. One such assumption is that the manifestations of efficacy and toxicity observed *in vitro* are relevant when extrapolated to conditions *in vivo*. A second major assumption is that the study of single cells rather than organs, tissue cultures or micromass preparations can still provide usable information when making decisions related to the problems of medical, veterinary or environmental safety (Pearson, 1986). The use of single-celled organisms such as bacteria is very widespread, but yeasts, protista and trypsinised preparations of higher organisms from sponges to mammals have also been employed for testing *in vitro* (Turner, 1983). In addition, a wide range of isolated cells from humans has been used, with considerable ingenuity, for examining an extraordinarily wide range of pharmacological effects, and certain of these are considered below.

1. Red blood cells

(*a*) DRUG TRANSPORT

Studies of transport mechanisms of drugs across lipid membranes have been carried out as a model of absorption from the gut lumen (Taylor and Turner, 1981). This study showed that the red blood cell/plasma concentration ratios were more reliable than organic solvent partition ratios in predicting distribution of drugs *in vivo*.

(*b*) DEFORMABILITY

Studies of the effects of drugs on red cell deformability have been performed in order to provide information relevant to treatments for patients with Raynaud's disease and progressive systemic sclerosis (Sowemimo-Coker *et al.*, 1985). These investigations *in vitro* have led to clinical trials of calcium channel antagonists in these conditions. Subsequently, a product licence has been granted for one of these drugs, nifedepine, for the treatment of these hitherto intractable conditions.

(*c*) HAEMOLYSIS

Examination of the haemolytic effect of candidate molecules on red blood cells (Dolan, 1981) has been used to assess their efficacy as contraceptives. This is potentially a very important and useful technique in those parts of the world where strongly-held cultural beliefs prevent investigations of potentially useful synthetic and natural products on semen samples. The effects of drugs on the stabilization of the erythrocyte membrane against haemolysis have been used to assess their membrane-stabilizing action and hence their potential cardiotoxicity in overdose (Cassidy and Henry, 1986).

2. White blood cells

The multiplicity of receptors on white blood cells provides extensive opportunities for pharmacologists to examine fundamental processes affecting the regulation of receptor responsiveness and numbers. The few receptors not so far identified on human lymphocytes include the muscarinic cholinergic receptor and the dopamine receptor.

(*a*) LYMPHOCYTES

Lymphocytes have been isolated from human patients and volunteers to examine the biochemical pathology underlying anxiety (Lima and Turner, 1983), asthma (Conolly *et al.*, 1978), hypertension (Lima and Turner, 1982), and essential tremor (Kilfeather *et al.*, 1984).

(*b*) NEUTROPHILS

Neutrophils isolated from human blood have been employed in investigations of porphyrin-induced photodamage (Sandberg *et al.*, 1982).

3. Platelets

These apparently simple cells have monoamine uptake, storage and release properties which they share at least in part with cells in the central nervous system. Preparations of human platelets have been used to study the effects of neurotransmitter substances and drugs in depressive illnesses (Turner and Ehsanullah, 1977) and in Huntingdon's chorea. Isolated human platelets have also been studied in investigations of the effects of drugs on arachidonic acid metabolism (Ludere *et al.*, 1980) and of the relationship between aging (Davis and Silski, 1987), exercise (Lockette *et al.*, 1987) and the responsiveness of the sympathetic nervous system.

4. Reproductive tract

(*a*) HUMAN SPERMATOZOA

The development of simple, inexpensive techniques (Hong *et al.*, 1981; Raoof *et al.*, 1987; Chijioke *et al.*, 1988) to compare the effects of drugs on sperm motility, has led to the screening of a large number of drugs in order to find new fertility-enhancing agents (Jiang *et al.*, 1984) as well as new spermicides (Louis and Pearson, 1985). The results obtained correlate well with those obtained from use of other techniques for testing the efficacy of spermicides (Chijioke *et al.*, 1986). Examination of the relation between the lipophilicity and membrane-stabilizing properties of a series of drugs that reduce sperm motility (Hong and Turner, 1982) has led to a clearer understanding of the cardiotoxicity of antidepressant and opiate drugs in overdose (Zaman *et al.*, 1984; Cassidy and Henry, 1986; Henry and Cassidy, 1986; Cassidy and Pearson, 1986).

(*b*) HUMAN LUTEAL CELLS

The mechanism of the effect of chorionic gonadotrophin on progesterone accumulation has been studied in isolated human luteal cells (Thibier *et al.*, 1980) isolated from corpora lutea removed intact from the ovary during elective laparotomy.

5. Gastro-intestinal tract cells

Gastric mucosal cells obtained from tissue removed at gastric resection have been used to study the effects of histamine H_1 and H_2 receptor agonist and blocking drugs in investigations of new anti-peptic ulcer agents (Simon *et al.*, 1978; Gustavsson *et al.*, 1985) and in investigations of secretion of intrinsic factor (Schepp *et al.*, 1984).

6. Nervous system

The metabolic activity of human brain cells has been studied in isolated nerve ending preparations recovered from post-mortem material (Webster *et al.*, 1985).

7. Adipose cells

Glucose transport has frequently been examined in isolated human adipose cells, and in one such recent study (Karnieli *et al.*, 1986), adipose cells were obtained by biopsy of subcutaneous fat or omentum during elective surgery. These cells were then used to examine the mechanism of the effects of insulin on the entry of glucose into cells by measuring the transport of 3-O-methylglucose.

8. Muscle cells

Isolated muscle cells from the bronchi obtained at thoracotomy (Marthan *et al.*, 1987), and cardiac muscle obtained during open heart surgery (Mitchell *et al.*, 1986) have been used to examine the effects of noradrenaline and acetylcholine, respectively.

9. Respiratory tract

Ciliary epithelium from the nose obtained by nasal brushing (Rutland *et al.*, 1983) and isolated parenchymal lung tissue (Conolly and Greenacre, 1977) have been used to study the cells of the human respiratory tract.

Acknowledgement

The support of the Peel Medical Research and Lawson Tait Medical and Scientific Research Trust, and Family Health International is gratefully acknowledged.

References

Cassidy, S. L. and Henry, J. A. (1986). Rapid *in vitro* techniques for the assessment of the membrane-stabilising activity of drugs. *Fd Chem. Toxicol.*, Nos. 6/7, 807–9.

Cassidy, S. L. and Pearson, R. M. (1986). Effects of trazodone and nadolol upon human sperm motility. *Br. J. clin. Pharmacol.*, **22**, 119–21.

Chijioke, P. C., Zaman, S. and Pearson, R. M. (1986). Comparison of the potency of *d*-propranolol, chlorhexidine and nonoxynol-9 in the Sander–Cramer test. *Contraception*, **34**, 207–11.

Chijioke, P. C., Crocker, P. R., Gilliam, M., Owens, M. D. and Pearson, R. M. (1988). Importance of filter structure for trans-membrane migration studies of sperm motility. *Hum. Reprod.*, **3**, 241–44.

Conolly, M. E. and Greenacre, J. K. (1977). The beta-adrenoceptor of the human lymphocyte and human lung parenchyma. *Br. J. Pharmacol.*, **59**, 17–23.

Conolly, M. E., Schofield, P. and Greenacre, J. K. (1978). Desensitisation of the beta-adrenoceptor of lymphocytes from normal subjects and asthmatic patients *in vitro*. *Br. J. clin. Pharmacol.*, **5**, 199–206.

Davies, P. B. and Silski, L. (1987). Ageing and the alpha-2 adrenergic system of the platelet. *Clin. Sci.*, **73**, 507–13.

Dolan, M. M. (1981). *In vitro* method for evaluation of spermicides by haemolytic potency. *Fert. Steril*, **36**, 248–9.

Gustavsscn, S., Mardh, S., Norberg, L., Nyren, O. and Wollert, H. (1985). Omeprazole, cimetidine and ranitidine: inhibition of acid production in isolated human parietal cells. *Scand. J. Gastroenterol.*, **20**, 917–21.

Henry, J. A. and Cassidy, S. L. (1986). Membrane stabilising activity; a major cause of fatal poisoning. *Lancet*, **i**, 1414–17.

Hong, C. Y., Chaput de Saintonge, D. M. and Turner, P. (1981). A simple method to measure drug effects on human sperm motility. *Br. J. clin. Pharmacol.*, **11**, 751–3.

Hong, C. Y. and Turner, P. (1982). Influence of lipid solubility on the sperm immobilising effect of beta-adrenoceptor blocking drugs. *Br. J. clin. Pharmacol.*, **14**, 269–72.

Jiang, C. S., Kilfeather, S. A., Pearson, R. M. and Turner, P. (1984). The stimulatory effects of caffeine, theophylline, lysine and 3-isobutyl-l-methylxanthine on human sperm motility. *Br. J. clin. Pharmacol.*, **18**, 258–62.

Karnieli, E., Chernow, B., Hissin, P. J., Simpson, I. A. and Foley, J. E. (1986). Insulin stimulates glucose transport in isolated human adipose cells through a translocation of intracellular glucose transporters to the plasma membrane: a preliminary report. *Hormone metabol. Res.*, **18**, 867–8.

Kilfeather, S. A., Maserella, A., Findley, L. J. and Turner, P. (1984). Beta-adrenoceptor involvement in tremor production: possible defects in essential tremor. In *Movement Disorders – Tremor*. Eds Findley, L. J. and Capildeo, R. pp. 225–45. London, Macmillan.

Lima, D. A. and Turner, P. (1982). Beta-blocking drugs increase responsiveness to prostacyclin in hypertensive patients. *Lancet*, **ii**, 444.

Lima, D. A. and Turner, P. (1983). Propranolol increases reduced beta-receptor function in severely anxious patients. *Lancet*, **ii**, 1505.

Lockette, W., McCurdy, R., Smith, S. and Carrettero, O. (1987). Endurance running and human alpha-2 adrenergic receptors on platelets. *Med. Sci. Sports Exerc.*, **19**, 7–10.

Louis, S. M. and Pearson, R. M. (1985). A comparison of the effects of nonoxynol-9 and chlorhexidine on sperm motility. *Contraception*, **32**, 199–205.

Ludere, J. R., Demers, L. M., Janson, R. W., Nomides, C. T. and Hayes, A. H. (1980). Effect of hydralazine on arachidonic acid metabolism in isolated, washed human platelets. *Res. Commun. Chem. Pathol. Pharmacol.* **28**, 43–52.

Marthan, R., Savineau, J-P. and Mironneau, J. (1987). Acetylcholine-induced contraction in human isolated bronchial smooth muscle: role of an intracellular calcium store. *Respir. Physiol.*, **67**, 127–35.

Mitchell, M. R., Powell, T., Sturridge, M. F., Terrar, D. A. and Twist, V. W. (1986). Electrical properties and response to noradrenaline of individual heart cells isolated from human ventricular tissue. *Cardiovasc. Res.*, **20**, 869–76.

Pearson, R. M. (1986). *In-vitro* techniques: can they replace animal testing? *Human Reprod.*, **1**, 559–60.

Raoof, N. T., Pearson, R. M. and Turner, P. (1987). A modified transmembrane method for measuring the effect of drugs on sperm motility. *Br. J. clin. Pharmacol.*, **24**, 319–21.

Rutland, J., Penketh, A., Griffin, W. M., Hodson, M. E., Batten, J. C., Sandberg, J. and Cole, P. J. (1983). Cystic fibrosis serum does not inhibit human ciliary beat frequency. *Am. Rev. resp. Dis.*, **128**, 1030–4.

Sandberg, S., Glette, J., Hopen, G., Solberg, C. O. and Romslo, I. (1982). Porphyrin-induced photodamage to isolated human neutrophils. *Photochem. Photobiol.*, **34**, 471–5.

Schepp, W., Miederer, S. E. and Ruoff, H. J. (1984). Intrinsic factor secretion from isolated human gastric mucosa cells. *Biochim. Biophys. Acta*, **804**, 192–9.

Simon, B., Kather, H. and Kommerell, B. (1978). Histamine sensitive adenylate cyclase of human gastric mucosa: a model for H_2 receptor excitation. *Br. J. clin. Pharmacol.*, **5**, 277–8.

Sowemimo-Coker, S. O., Debbas, N. M. G., Kovacs, I. B. and Turner, P. (1985). *Ex-vivo* effects of nifedipine, nisoldepine and nitrendepine on filterability of red blood cells from healthy volunteers. *Br. J. clin. Pharmacol.*, **20**, 152–4.

Taylor, E. A. and Turner, P. (1981). The distribution of propranolol, pindolol and atenolol between human erythrocytes and plasma. *Br. J. clin. Pharmacol.*, **12**, 543–8.

Thibier, M., El Hassan, N., Clark, M. R., LeMaire, E. J. and Marsh, J. M. (1980). Inhibition by estradiol of human chorionic gonadotrophin-induced progesterone accumulation in isolated human luteal cells; lack of mediation by prostaglandin F. *J. clin. Endocr. Metab.*, **50**, 590–2.

Turner, P. (Ed.). (1983). *Animals in Scientific Research: An Effective Substitute for Man?* London, Macmillan Press.

Turner, P. and Ehsanullah, R. S. B. (1977). Human platelet uptake of 5-hydroxy tryptamine and dopamine *in vitro*; what relevance to their antidepressive and other central actions? *Postgrad. Med. J.*, **32** (Suppl. 4), 14–17.

Wester, P., Bateman, D. E., Dodd, P. R., Edwardson, J. A., Hardy, J. A., Kidd, A. M., Perry, R. H. and Singh, G. B. (1985). Agonal status affects the metabolic activity of nerve endings isolated from postmortem human brain. *Neurochem. Pathol.*, **3**, 169–80.

Zaman, S., Lamb, J. M., Esberger, D. A. and Pearson, R. M. (1984). Inhibition of sperm motility by opiate drugs. *Br. J. clin. Pharmacol.*, **18**, 320P.

CHAPTER 2

The use of human cell cultures in the development of anti-rheumatic drugs

G. P. Lewis

Introduction

A method has been used in which the effect of anti-inflammatory and anti-rheumatic drugs have been examined on human peripheral blood mononuclear cell interactions *in vitro* (Gordon and Lewis, 1984).

Blood was collected from donors and the mononuclear cells isolated and set up in culture. When necessary, non-adherent lymphocytes were separated from the adherent macrophages and cultured separately or the whole population of mononuclear cells can be cultured together.

When mononuclear cells are stimulated with antigen or mitogen, there is an interaction between lymphocytes and macrophages (Fig. 2.1). Lymphokines are released from lymphocytes and one of these—(MAF) (possibly γ-interferon)—activates macrophages. These cells in turn produce interleukin-l (IL-1) which acts on lymphocytes, converting some of the T-cells into cells which are capable of producing interleukin-2 (IL-2) or T-cell growth factor. In order for IL-2 to fulfil its role, it is necessary for other T-cells to be stimulated by the mitogen to become IL-2 responder cells. This implies the induction of IL-2 receptors on the cells.

Thus, when we measure T-cell proliferation by estimating labelled thymidine (^{3}H-TdR) incorporation in a population of mononuclear cells, we are measuring the end product of a complicated series of events which are dependent upon interleukins 1 and 2. These events are of course central to the problem of chronic inflammation, where T-cell proliferation represents part of the ongoing nature of the disease.

Interleukin-1 not only induces the maturation of T-cells to become IL-2 producer cells, but has other inflammatory actions as well. Among them is stimulation of fibroblasts and rheumatoid synovial cells to produce prostaglandins (mainly PGE_2) and neutral proteinases such as collagenase, a method which can be used for the assay of IL-1 production.

The effect of anti-inflammatory agents and the anti-rheumatic compounds has been examined in this system.

Methods

MATERIALS

Cytokine preparations were purchased from the sources indicated: interleukin-1 (IL-1; Genzyme, UK; 100 U/ml), γ-interferon (γ-IF; Ventrex, USA; 100 U/ml), T-cell growth factor (Associated Biomedics Systems Inc. USA). γ-Interferon in the culture supernatant was measured directly after 72 h using a Centor RIA kit.

Auranofin (AF; Smith, Kline and French, Ltd) was dissolved in absolute alcohol at 20 mg/ml and further diluted in Dulbecco's modified Eagles' medium (DMEM, Gibco) to the required concentrations. Aurothiomalate (ATM; 10% Myocrisin, May and Baker Ltd) was diluted in DMEM as required. Aurothioglucose (ATG; Sigma) was dissolved at 8 mg/ml in DMEM and sterilized by membrane-filtration (0.22 μm Millex) prior to dilution in DMEM. Dexamethasone sodium phosphate (Merck, Sharp and Dohme), mepacrine dihydrochloride (Boots) and chloroquine diphosphate (Sigma) were dissolved directly in DMEM. Azathioprine and 6-mercaptopurine (both Sigma) were dissolved in dilute NaOH and further diluted in DMEM. Cyclosporin-A (Sandoz) was dissolved in

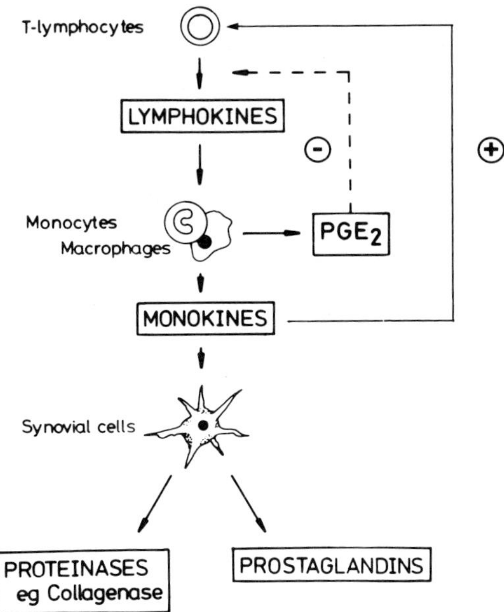

Fig. 2.1 Activated lymphocytes release lymphokines, at least one of which activates monocyte/macrophages which, in turn, release monokines which amplify lymphocyte proliferation. The macrophages also release prostaglandins which down-regulate the activated lymphocytes. The monokines also release proteinases, such as collagenase, and prostaglandins from fibroblasts and synovial lining cells

ethanol–Tween 80 and diluted in DMEM (Gordon and Nouri, 1981). Aqueous solutions were sterilized by membrane filtration (0.2 μm Millex) prior to further dilution in DMEM. Piroxicam (Pfizer) was dissolved in acetone; indomethacin (Merck, Sharp and Dohme) and PGE_2 (Sigma) were dissolved in ethanol; further dilutions were made in DMEM without filtration.

ISOLATION AND CULTURE OF MONONUCLEAR CELLS

Mononuclear cells were separated from defibrinated blood of normal volunteers by discontinuous gradient centrifugation using Ficoll-Paque (Pharmacia) (Gordon and Nouri, 1981). The washed cells were resuspended in Dulbecco's modified Eagles' medium (DMEM, Gibco) containing antibiotics (100 units penicillin/ml, 100 μg streptomycin/ml) and supplemented with 10% heat-inactivated (30 min, 56°C) foetal calf serum (HIFCS, Gibco). Aliquots of cell suspension containing 2×10^5 viable cells were dispensed into flat-bottomed microtitre wells for assessment of lymphocyte proliferative responses by ^{3}H-TdR incorporation (Gordon and Nouri, 1981); aliquots containing 5×10^5 viable cells were dispensed into multi-well plates (1.5 cm, Linbro) for measurement of PGE_2, IL-1 and IL-2 release. Drugs and mitogens were added at the beginning of culture and the volume was adjusted to give a final cell concentration of 1×10^6 viable cells/ml. Cell-free supernatant solutions collected after 48 h incubation were used for determination of PGE_2, IL-1 and IL-2 production.

The PGE_2 content of the unextracted mononuclear cell culture supernatants was measured by radioimmunoassay (RIA), using the antiserum and procedure described by Jose *et al.* (1981). IL-2 activity was assessed by ^{3}H-TdR incorporation into phytohaemagglutinin (PHA)-activated mononuclear cells at a dilution of 1 in 5. IL-1 (mononuclear cell factors: MCF) activity was determined as PGE_2 production by rheumatoid synovial cells at a dilution of 1 in 10. With a view to the detection of IL-2-like activity and the determination of drug effects on IL-2-induced ^{3}H-TdR uptake, IL-2 reactive lymphocytes were generated in 75 cm^2 tissue culture flasks (Falcon) by incubation of 1×10^6 mononuclear cells/ml DMEM + 10% HIFCS with PHA (1 μg/ml) for 7 days at 37°C in a 5% CO_2 atmosphere. These cells were maintained for a further 1–2 weeks in DMEM + 10% HIFCS containing lectin-free IL-2 (1 U/ml, Bethesda Research Laboratories) prior to use. IL-2-induced proliferation was assessed using 1×10^6 cells/ml, on the basis of the ^{3}H-TdR incorporation during the subsequent 72 h culture, essentially as for freshly isolated mononuclear cells. The drugs were added at the beginning of culture and their effects were examined using 1 U/IL-2/ml in freshly isolated mononuclear cells. Adherent monocytes were prepared by incubating for 2 h at 37°C and non-adherent cells were removed by repeated washing. Drugs were then added and incubation continued for a further 48 h.

ISOLATION AND CULTURE OF SYNOVIAL CELLS

Adherent synovial cell cultures were established essentially as described by other workers (Dayer *et al.*, 1976), using synovectomy material obtained from rheumatoid patients undergoing corrective surgery. Adherent synovial cells (3×10^5/ml), maintained in DMEM supplemented with 10% heat-inactivated foetal calf serum (HIFCS) release large amounts of prostaglandin E_2 and collagenase during primary culture. Upon sub-culturing, the production of mediators declines to low or undetectable levels. These quiescent cells were used for assay of IL-1 action through stimulation of PGE_2 release. After 48 h incubation periods at 37°C in 5% CO_2, the cell-free supernatants were removed and the PGE_2 content assayed by RIA. Mononuclear cell supernatants were assayed on synovial cells at 1 in 10 and 1 in 25 dilutions. At these dilutions, the direct effect of drugs on PGE_2 production by mononuclear cells was insignificant.

In collaboration with Dr W. Harvey (Eastman Dental Hospital, London), synovial cell supernatants were assayed for collagenase by digestion of ^{3}H-acetylated rat skin collagen fibrils, using trypsin to activate latent enzyme.

Results

NON-STEROID ANTI-INFLAMMATORY DRUGS

Prostaglandin (PG) E_2 is not only produced by synovial cells but by macrophages as well, which raises the question of the importance of cyclo-oxygenase inhibitors on mononuclear cell populations (Lewis, 1983).

Several cyclo-oxygenase inhibitors have been examined but indomethacin has been

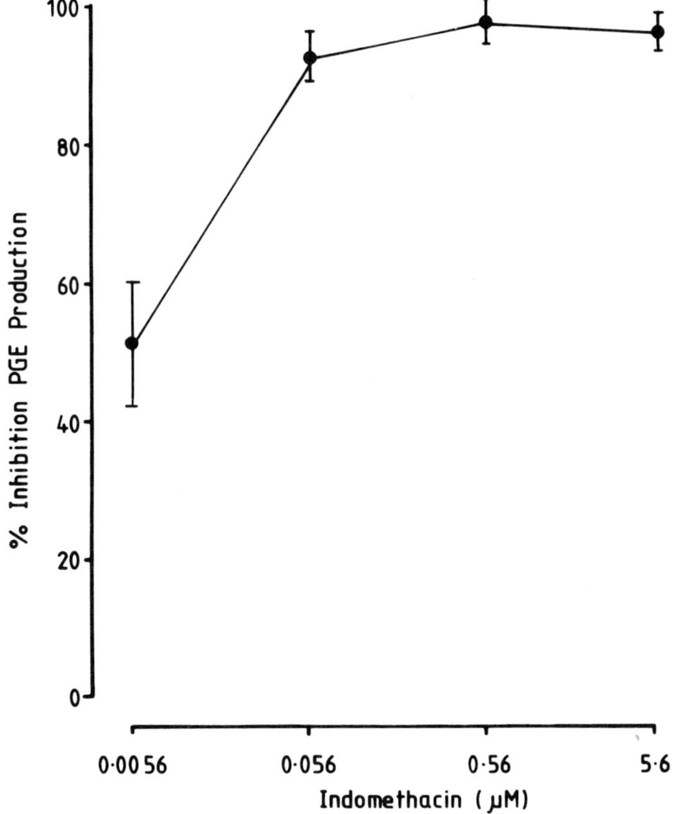

Fig. 2.2 Inhibition by indomethacin of PGE_2 production by blood mononuclear cells. Each value is the mean of 7 replicate cultures; vertical lines indicate SEM

taken as an example. Indomethacin was found to inhibit prostaglandin (PGE_2) production by PHA-stimulated mononuclear cells in a concentration range of 5.6–560 nM. As shown in Fig. 2.2, the IC_{50} was 5.6 nM and the near maximum effect was produced by 560 nM (Gordon and Lewis, 1984).

Within this concentration range, indomethacin caused a dose-related enhancement of ^{3}H-TdR incorporation as shown in Fig. 2.3. However, this enhancement of T-lymphocyte proliferation only occurred at sub-optimal concentrations of PHA, even when a supramaximal concentration of indomethacin was used (Fig. 2.4). In order to examine whether this effect of indomethacin was mediated by inhibition of PG production, experiments using exogenous PGE_2 were performed. Figure 2.5 shows that the enhancement of sub-optimal PHA stimulation caused by indomethacin was overcome by the addition of PGE_2 (3–30 ng/ml). Such concentrations are within the range of PG levels formed by mononuclear cells stimulated with mitogen *in vitro*, or formed locally *in vivo*. This endogenous PG would therefore be available to fulfil the role of mediator of a down-regulating mechanism.

The exogenous PG had little or no effect on ^{3}H-TdR uptake in the indomethacin-free control cultures in which comparable amounts of PGE_2 were released. It seems likely that the PGE_2 already formed in the mononuclear cell culture inhibited proliferation

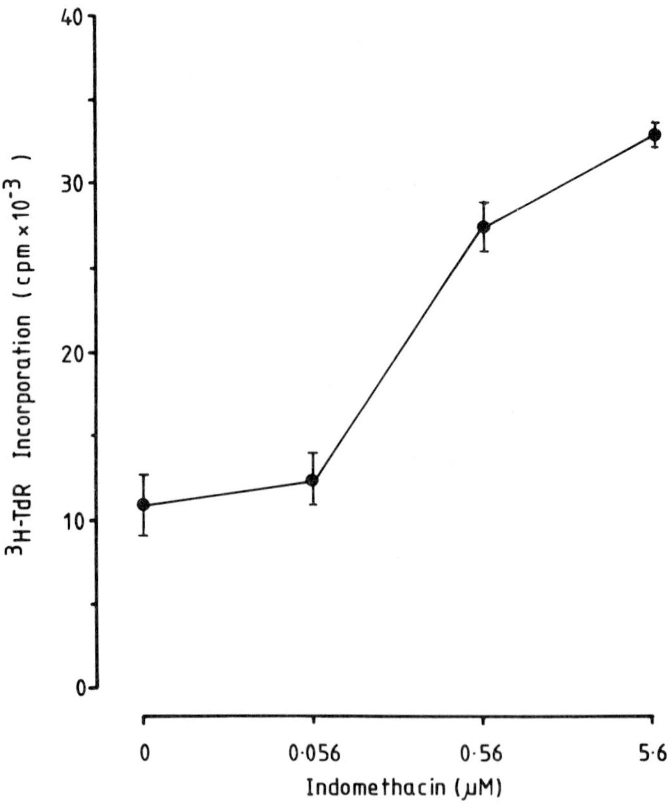

Fig. 2.3 Enhancement by indomethacin of sub-optimal (0.03 μg/ml) PHA-induced ^{3}H-TdR incorporation. Each value is the mean of 5 replicate cultures; vertical lines indicate SEM

maximally, so that addition of exogenous PGE_2 was unable to increase the effect. A similar explanation was given by Gordon *et al.* (1979) for the greater susceptibility of indomethacin-treated mononuclear cells compared with untreated cells. Furthermore, they suggested that the presence of PGs in the serum added to cell culture media might well explain some of the variability in the response of such cells to PGs and to mitogen reported in various laboratories.

It was found that indomethacin and T-cell growth factor or interleukin-2 (IL-2) behave in the same way, in that they both selectively enhanced sub-optimal PHA-induced ^{3}H-TdR incorporation but did not affect optimal responses as shown in Fig. 2.6.

At this point the question arises—is IL-2 involved in the action of cyclo-oxygenase inhibitors (Lewis and Barrett, 1986)? There are at least three ways in which indomethacin could enhance T-cell proliferation and probably other T-cell functions via IL-2:

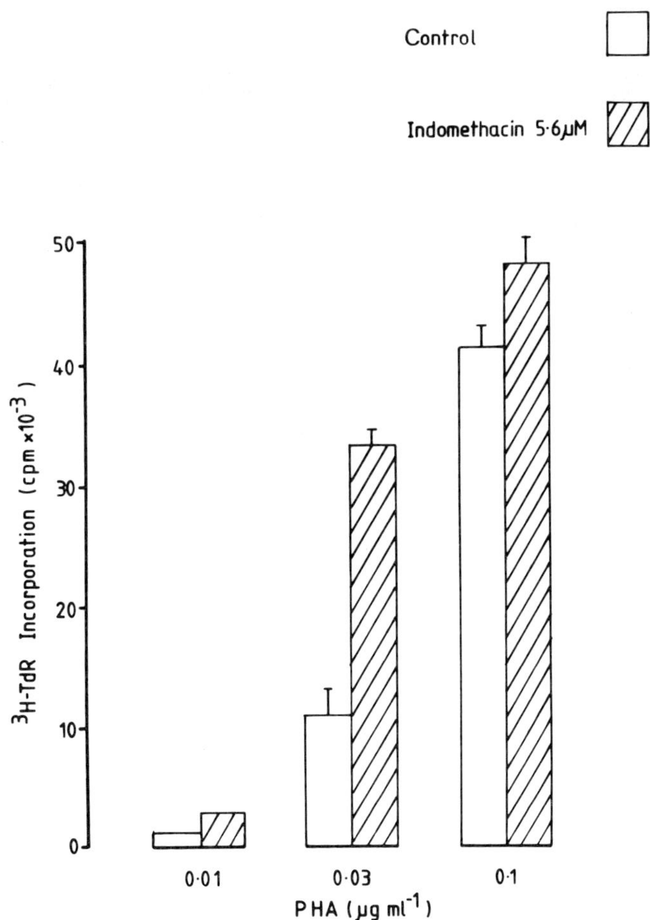

Fig. 2.4 Effect of indomethacin (5.6 μM) compared with untreated controls on PHA-stimulated ^{3}H-TdR incorporation into blood mononuclear cells

(1) enhancement of the action of IL-2;
(2) enhancement of the induction of IL-2 receptors on IL-2 responder cells; and
(3) increased IL-2 production by IL-2 producer cells.

In examining the possibility (1), it was found that indomethacin neither enhanced nor inhibited IL-2-induced proliferation of IL-2 responder T-cells, as shown in Fig. 2.7.

On the other hand, Fig. 2.7 shows that both PGE_2 and dexamethasone directly inhibited the action of IL-2. Furthermore, the concentration of PGE_2 necessary for this effect was well within the range of concentrations produced by a population of activated mononuclear cells, although inhibition did not exceed about 40%.

The failure of indomethacin to enhance IL-2-induced responses reflects the lack of endogenous PGE_2 production by the IL-2 responder T-cells. Our experiments confirm earlier findings in mice (Baker *et al.*, 1981) that as little as 1–1000 nM PGE_2 and PGE_1 inhibited IL-2-dependent T-cell proliferation by about 40%.

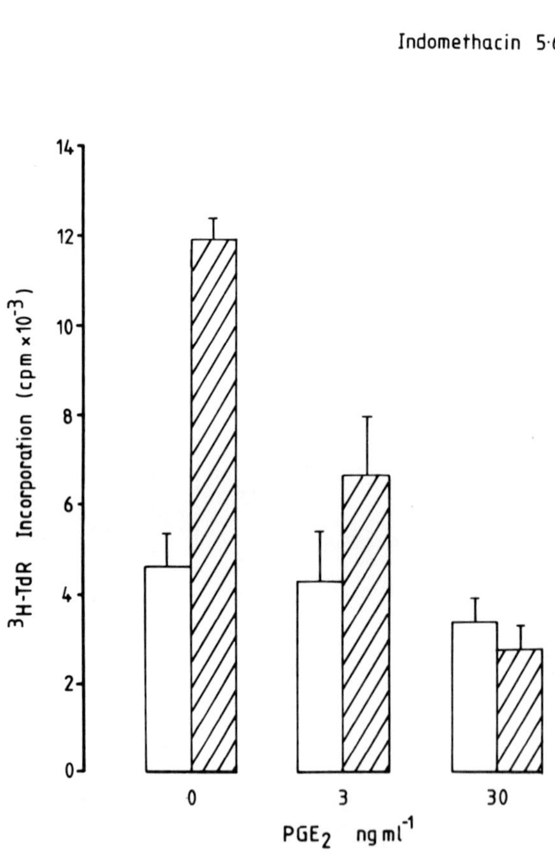

Fig. 2.5 Effect of exogenous PGE_2 on indomethacin (5.6 μM)-enhancement of sub-optimal (0.03 μg/ml) PHA-induced ^{3}H-TdR incorporation into blood mononuclear cells, compared with results for untreated control cultures. Each value is the mean $\pm$ SEM for 5 replicate cultures

Possibility (2) has been examined by Walker *et al.* (1983) in human peripheral blood leukocytes and it was found that, over a concentration range of 0.1 to 1000 nM, PGE_2 had no effect on the formation of IL-2 receptors.

Several investigations have been carried out into possibility (3) in mice (Baker *et al.*, 1981) and human (Rappaport and Dodge, 1982; Walker *et al.*, 1983; Gordon and Lewis, 1984; Chouaib *et al.*, 1985) leukocytes. All the investigators agree that PGE_2 inhibits IL-2 production, usually in about the same concentration as it has been reported to inhibit IL-2 action. Furthermore, indomethacin enhances the production of IL-2 in mitogen-stimulated blood mononuclear cells, as shown in Fig. 2.8.

Thus, the PGE-mediated suppression of mitogen-induced T-cell proliferation can be attributed to an inhibition of both production and action of the T-cell specific mitogen, IL-2.

Besides the capacity to control lymphocyte activation by a direct action, it has been shown (Webb *et al.*, 1979; Webb and Nowowiejski, 1981) that PGE_2 can activate T-

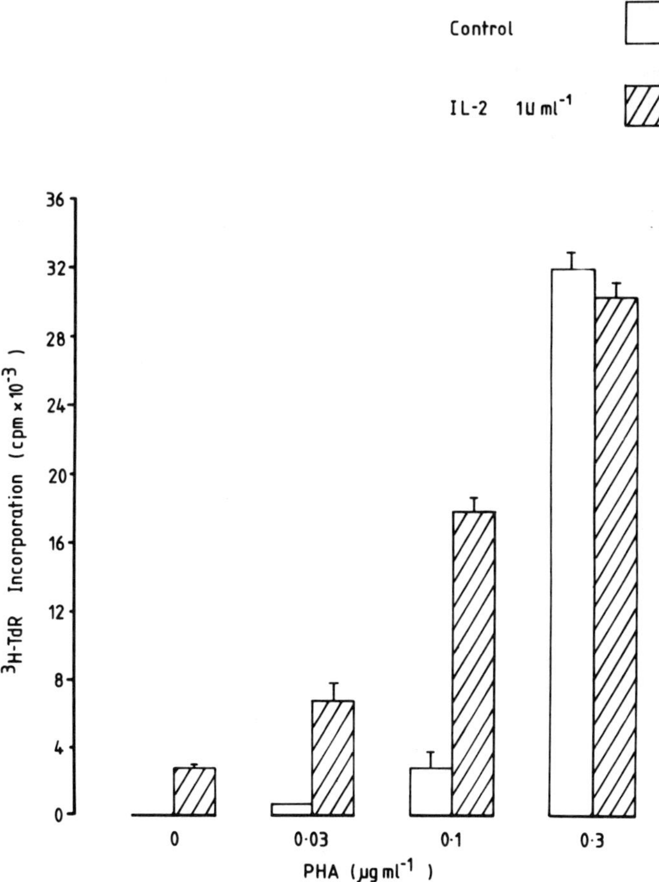

Fig. 2.6 Effect of exogenous IL-2 (1 U/ml) on PHA-stimulated ^{3}H-TdR incorporation into blood mononuclear cells, compared with results for untreated control cultures. Each value is the mean $\pm$ SEM of 5 replicate cultures

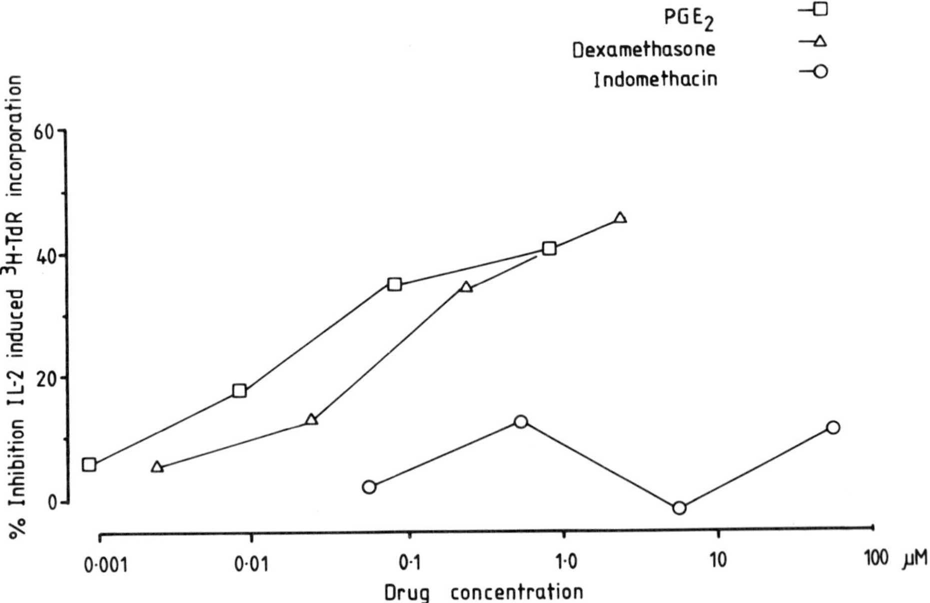

Fig. 2.7 Effects of indomethacin, PGE_2 and dexamethasone on IL-2-induced ^{3}H-TdR incorporation by PHA-activated blood lymphocytes. Each value is the mean of 6 replicate cultures. Control ^{3}H-TdR incorporation was 15 711–32 511 cpm

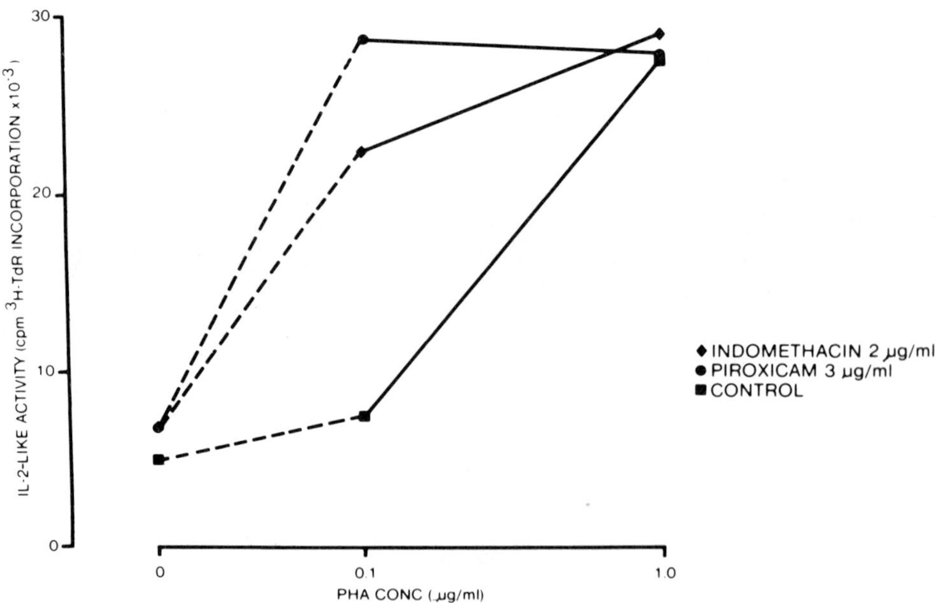

Fig. 2.8 Effect of indomethacin (5.6 μM) on the release of IL-2-like activity by PHA-activated blood mononuclear cells, compared with control cultures. Each value is the mean of 6 replicate cultures

suppressor cells. Murine glass adherent T-cells were found to suppress DNA synthesis in a PHA-stimulated non-adherent T-lymphocyte population. Furthermore, when the total cell populations were exposed to PGE_2 (10^{-5} M) they were more suppressive than when exposed to PHA. Finally, the same authors showed that PGE_2 stimulated the adherent T-cells to release a soluble suppressor factor which appeared to be a low-molecular-weight peptide. Similar findings have been made in human blood.

ANTI-RHEUMATIC DRUGS

In contrast to the cyclo-oxygenase inhibitors which enhance lymphocyte proliferation, the other classes of anti-rheumatic drugs so far tested, including antimalarials, immunosuppressive agents, glucocorticoids, cyclosporin-A and gold compounds were shown to exert powerful anti-proliferative activities.

Anti-malarial drugs

As shown in Fig. 2.9, mepacrine (2–6 μM, 1–3 μg/ml) and chloroquine (6–60 μM, 3–30 μg/ml) both inhibited optimal PHA-induced ^{3}H-TdR incorporation in a dose-related manner, with very steep dose–inhibition curves. Figure 2.9 also illustrates the more gradual inhibition of IL-1 production by lower concentrations of mepacrine (0.6–6 μM, 0.3–3 μg/ml) and chloroquine (0.6–20 μM, 0.3–10 μg/ml). These effects were not attributed to interference from traces of drugs introduced to the synovial cell assay from the mononuclear cell cultures. There was no significant effect of mepacrine (0.2 μM) or chloroquine (0.6 μM) on IL-1-induced PGE_2 production by synovial cells whereas, at

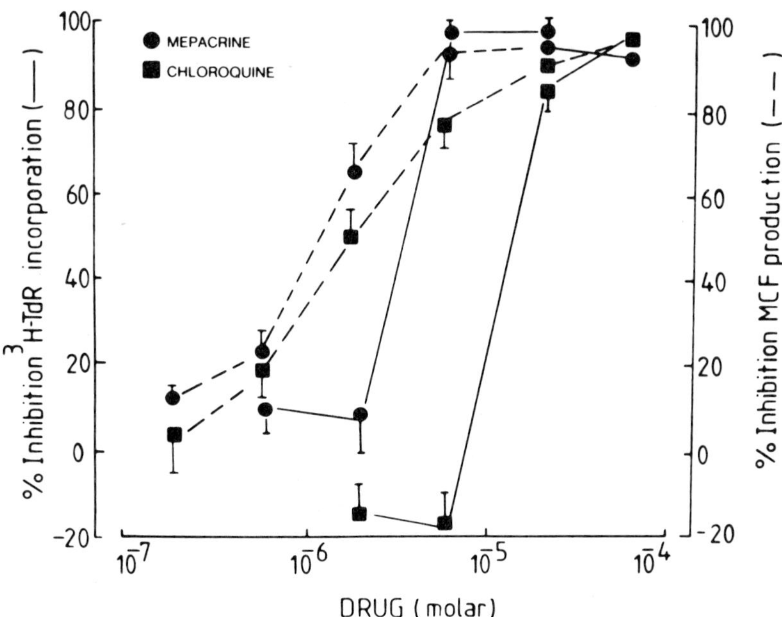

Fig. 2.9 Inhibition by mepacrine and chloroquine of optimal (1 μg/ml) PHA-stimulated ^{3}H-TdR incorporation (———) and IL-1 (MCF) release (– – –) by blood mononuclear cells. Each value is the mean $\pm$ SEM of 8 replicate cultures from two separate experiments

equivalent concentrations (i.e. 10-fold higher before dilution), approximately 70% inhibition of IL-1 release from mononuclear cells was observed. Nevertheless, inhibition of IL-1 release, IL-1 action and lymphocyte proliferation were found to be overlapping effects; this was particularly true for mepacrine. Inhibition of IL-2-induced ^{3}H-TdR incorporation by PHA-activated lymphocytes was also found at similar concentrations.

Immunosuppressive agents

Figure 2.10 illustrates the dose-related inhibition by cyclosporin-A (0.01–1.0 μM, 0.012–1.2 μg/ml) azathioprine (1–100 μM, 0.28–28 μg/ml) and 6-mercaptopurine (0.1–100 μM, 0.015–15 μg/ml) of optimal PHA-stimulated ^{3}H-TdR incorporation. In contrast to the anti-malarials, all three immunosuppressive drugs were less effective in inhibiting IL-1 production than lymphocyte proliferation, although parallel dose–inhibition curves were given by the individual agents for both effects. As with the anti-malarials, their effects on IL-1 release were not attributable to interference with IL-1 action.

Gold compounds

A somewhat more complex study has been made of the gold compounds (Gordon and Lewis, 1984; Barrett and Lewis, 1986).

Three gold compounds, auranofin (AF; 0.2–3.1 μg/ml, 0.3–5.0 μM), sodium aurothiomalate (ATM; 50–800 μg/ml, 0.125–2.0 mM) and aurothioglucose (ATG; 50–800 μg/ml, 0.125–2.0 mM) were shown to inhibit both sub-optimal and optimal PHA-induced ^{3}H-TdR incorporation (Fig. 2.11) in a dose-related manner. AF (IC_{50}

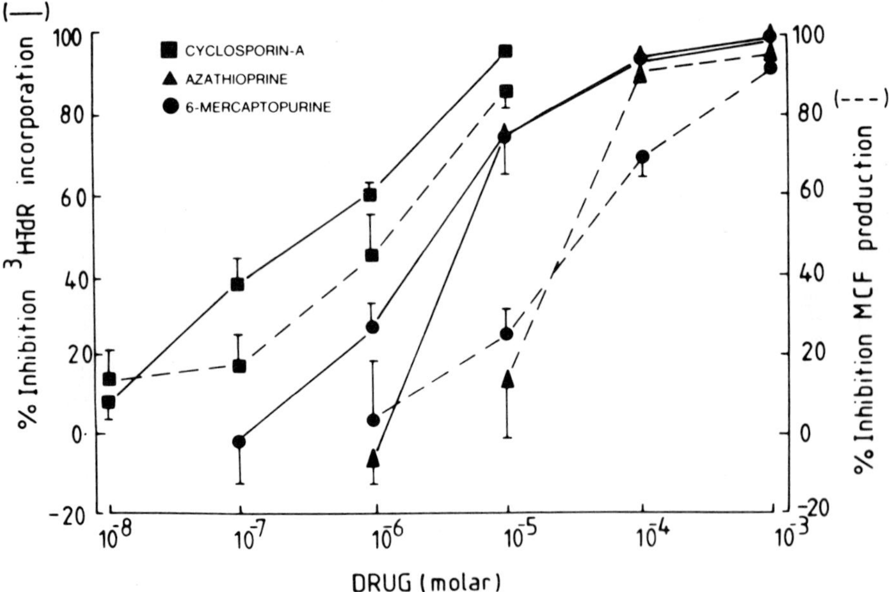

Fig. 2.10 Inhibition by cyclosporin-A, azathioprine and 6-mercaptopurine of optimal (1 μg/ml) PHA-stimulated ^{3}H-TdR incorporation (———) and IL-1 (MCF) release (– – –) by blood mononuclear cells. Each value is the mean ± SEM of 8 replicate cultures from two separate experiments

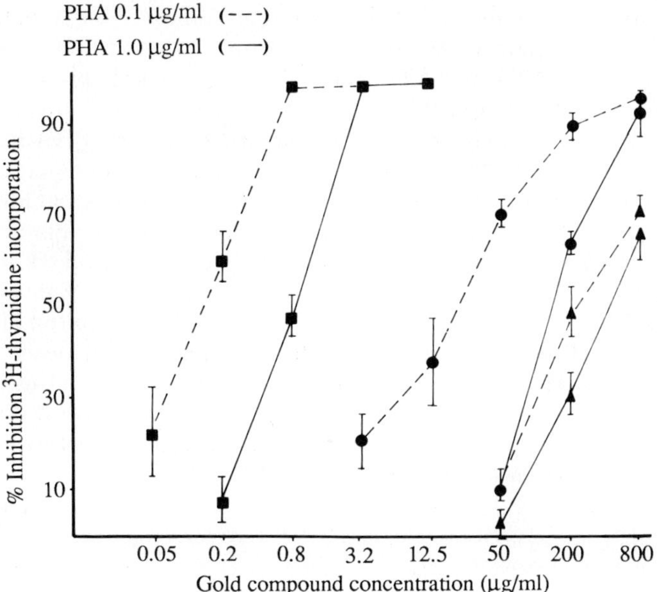

Fig. 2.11 Effects of auranofin (■), sodium aurothiomalate (●) and aurothioglucose (▲)on sub-optimal (0.1 μg/ml, – – –) and optimal (1.0 μg/ml, ———) PHA-induced ^{3}H-TdR incorporation by blood mononuclear cells. Each value is the mean ± SEM of 12 replicate cultures from three separate experiments

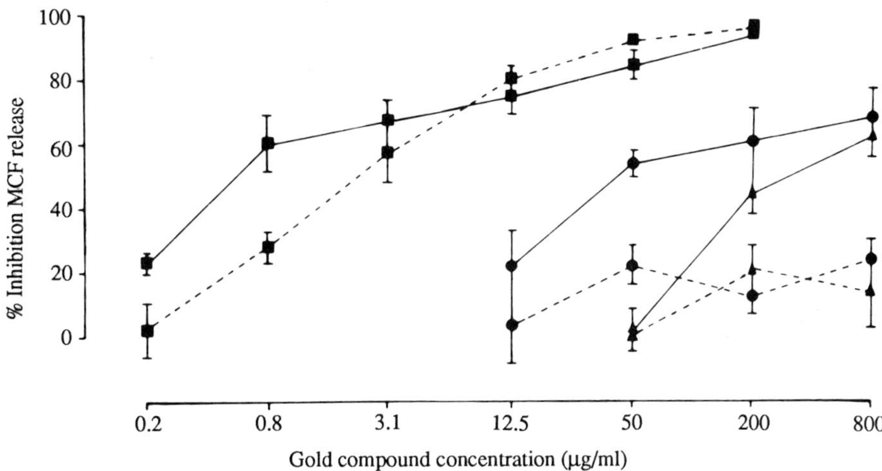

Fig. 2.12 Effects of auranofin (■), sodium aurothiomalate (●) and aurothioglucose (▲) on IL-1 release by unstimulated monocytes (– – –) and PHA-(1 μg/ml)-stimulated mononuclear cells (———). Each value is the mean ± SEM of 6–12 replicate cultures from two to three separate experiments

0.84 μg/ml) was the most potent inhibitor, being approximately 160 and 480 times more active on a weight basis than ATM (IC_{50} 138 μg/ml) and ATG (IC_{50} 405 μg/ml). As shown in Fig. 2.12 neither ATM nor ATG affected spontaneous IL-1 release by monocytes at concentrations up to 800 μg/ml. In addition, neither compound affected lipopolysaccharide (LPS)-induced IL-1 release (not illustrated). However, PHA-stimulated IL-1 production was inhibited in a dose-related manner by both compounds with IC_{50} values of 40 and 280 μg/ml, respectively. On the other hand, AF inhibited spontaneous and LPS-induced IL-1 release (IC_{50} 2.1 μg/ml) but, like the other gold compounds, was more effective in inhibiting PHA-stimulated IL-1 release (IC_{50} 0.54 μg/ml). Furthermore, AF was exceptional in inhibiting IL-1-induced PGE_2 synthesis by synovial cells (IC_{50} 0.84 μg/ml; Fig. 2.12). At a concentration of 800 μg/ml, ATM inhibited IL-1-stimulated PGE_2 production by 60% whereas ATG was ineffective (Fig. 2.13). However, the effects of the gold compounds on IL-1 release could not be accounted for in terms of interference with IL-1 activity by the amounts carried over in the mononuclear cell supernatant. AF not only inhibited IL-1-induced PGE_2 production but also the release of collagenase brought about by IL-1 as shown in Fig. 2.14.

Cytokines IL-2, γ-IF and IL-1 are established mediators of intercellular reactions between sub-populations of mononuclear cells. We were therefore interested in the effect of their addition upon the anti-proliferative effects of the gold compounds which would suggest their involvement in the mode of action of the compounds.

Exogenous IL-2 (1 U/ml) partly reversed the inhibitory effects of the gold salts on sub-optimal PHA-stimulated cells as illustrated in Fig. 2.15. Recombinant IL-2 (100 U/ml) had a similar effect (results not shown). The effect was notable against concentrations of gold salts which caused less than 70% inhibition of the proliferative response. However, IL-2 did not reverse inhibition of mononuclear cells stimulated with an optimal PHA concentration.

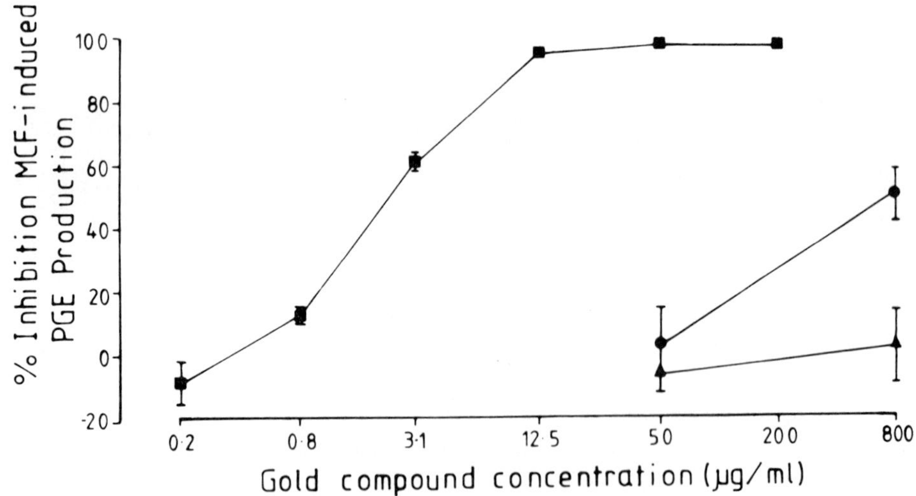

Fig. 2.13 Effects of auranofin (■), sodium aurothiomalate (●) and aurothioglucose (▲) on IL-1-induced PGE_2 production by human rheumatoid synovial cells. Each value is the mean ± SEM of 6–12 replicate cultures from two or three separate experiments

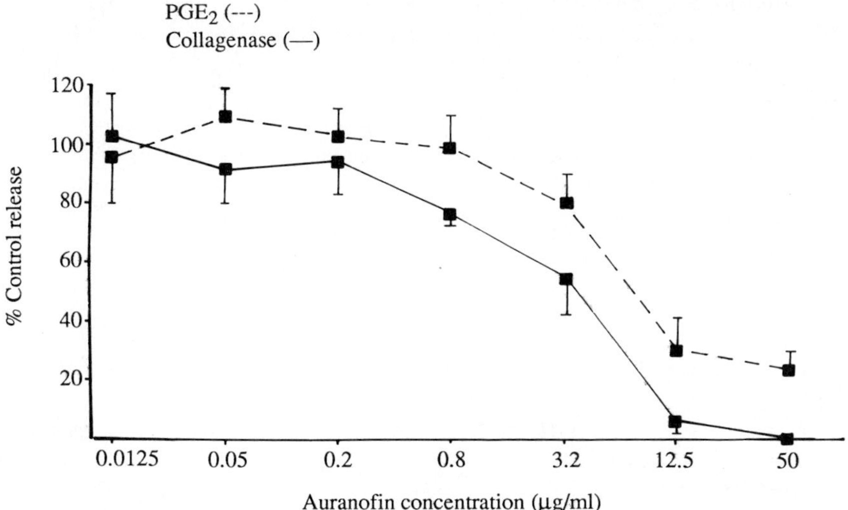

Fig. 2.14 Effects of auranofin upon PGE_2 release (– – –) and collagenase release (———) from primary culture rheumatoid synovial cells. Each value is the mean ± SEM of 3 replicates from one experiment

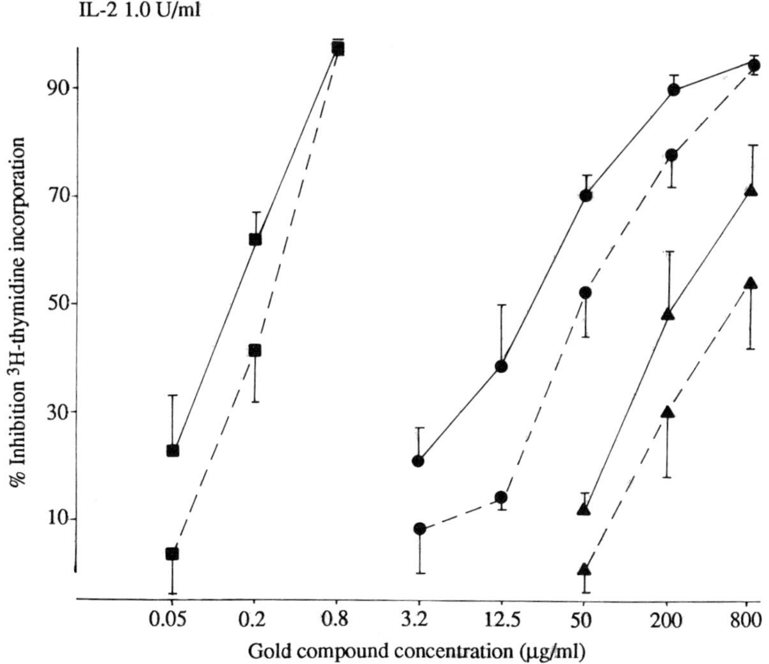

Fig. 2.15 Inhibition by auranofin (■), sodium aurothiomalate (●) and aurothioglucose (▲) of sub-optimal PHA (0.1 $\mu g/ml$)-stimulated ^3H-TdR incorporation by mononuclear cells in the absence of (———) and presence of (– – –) I1-2 (1 U/ml). Each value is the mean ± SEM of 16 replicate cultures from four separate experiments

As illustrated in Fig. 2.16, the addition of γ-IF (10 U/ml) partly reversed the anti-proliferative effects of ATM and ATG but had no effect upon the action of AF. Addition of IL-1 (5 U/ml) had no effect upon the inhibitory action of any of the gold compounds as depicted in Fig. 2.17.

Figure 2.18 illustrates the effects of the gold compounds upon IL-2-induced proliferation of PHA-blast cells. Effective concentration ranges were similar to those that inhibited PHA-induced mononuclear cell proliferation. γ-IF production was also inhibited in this concentration range as shown in Fig. 2.19.

Discussion

It has been established by many investigators that anti-inflammatory drugs like indomethacin inhibit prostaglandin formation and this effect undoubtedly contributes to the symptomatic benefit of arthritic patients treated with these drugs. In mononuclear cell populations, enhanced lymphocyte proliferation in response to sub-optimal concentrations of T-cell mitogen PHA was shown to be a consequence of the inhibition of PGE_2 formation. The mechanisms underlying such enhancement were analysed and shown to involve increased release of mitogenic lymphokine IL-2 and, to a lesser extent, enhanced activity of IL-2. Taken together with the results of earlier studies using guinea pig (Gordon *et al.*, 1976; Bray *et al.*, 1978), murine (Baker *et al.*, 1981) and human (Rappaport and Dodge, 1982) mononuclear cells, the present findings are consistent with the view that cyclo-oxygenase inhibitors prevent the negative-feedback suppressive effects of

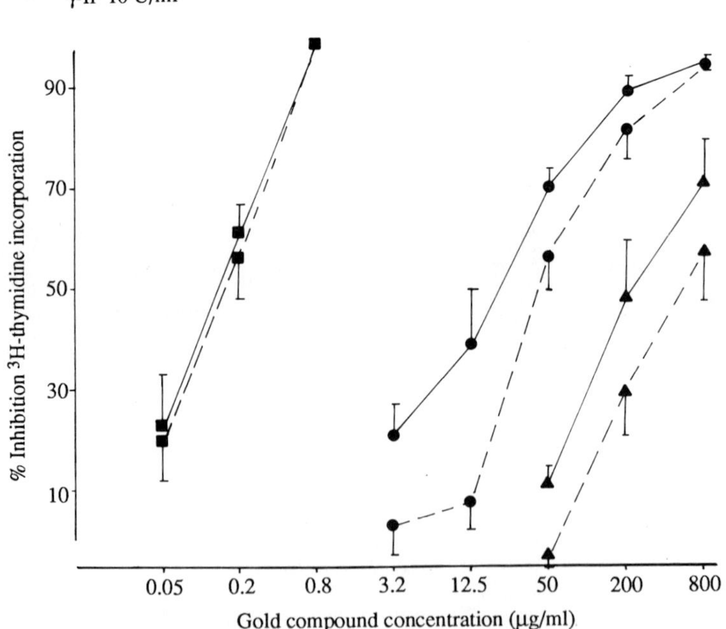

Fig. 2.16 Inhibition by auranofin (■), sodium aurothiomalate (●) and aurothioglucose (▲) of sub-optimal PHA (0.1 μg/ml)-stimulated ^{3}H-TdR incorporation by mononuclear cells in the absence of (———) and presence of (— — —) γ-interferon (γ-IF) (10 U/ml). Each value is the mean ± SEM of 16 replicate cultures from four separate experiments

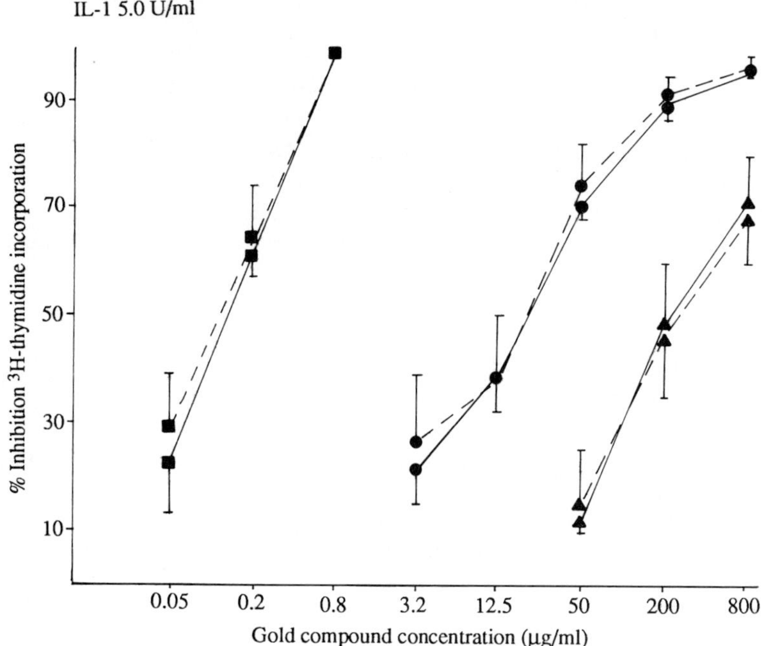

Fig. 2.17 Inhibition by auranofin (■), sodium aurothiomalate (●) and aurothioglucose (▲) of sub-optimal PHA (0.1 μg/ml)-stimulated ^{3}H-TdR incorporation by mononuclear cells in the absence of (———) and presence of (— — —) IL-1 (5.0 U/ml). Each value is the mean ± SEM of 16 replicate cultures from four separate experiments

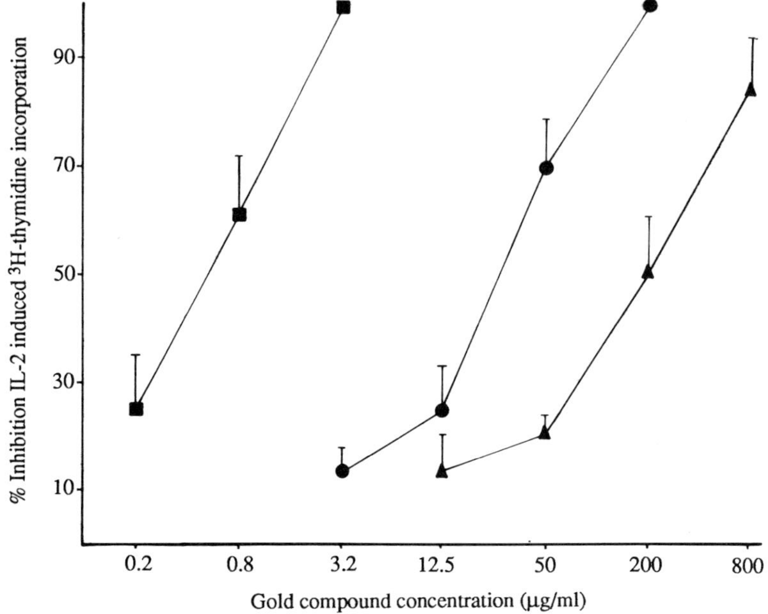

Fig. 2.18 Effects of auranofin (■), sodium aurothiomalate (●) and aurothioglucose (▲) on IL-2-induced ^{3}H-TdR incorporation by PHA-blast cells. Each value is the mean ± SEM of 9 replicate cultures from three separate experiments

monocyte/macrophage-derived PGE_2 on the secretion by T-lymphocytes of lymphokines such as mitogenic IL-2.

In addition to the prostaglandin-related suppressor system, there are at least two other suppressor systems in normal human blood that modulate T-cell lymphocyte proliferation and possibly B-lymphocyte activation, leading to immunoglobulin synthesis, as is found in the synovial membrane in rheumatoid arthritis (Rice *et al.*, 1978). An adherent cell suppressor system distinct from the PG-related cells can be demonstrated by increasing in cultures the concentration of cells that adhere to foreign surfaces. Furthermore, another distinct mitogen-induced suppressor system can be activated by a 40-h pre-incubation with concanavalin A (Con A) (Shou *et al.*, 1976).

The relative importance of these systems is difficult to assess at present but their clinical relevance is apparent from several findings. For example, serum from some patients with systemic lupus erythematosus was found to inhibit the mitogen-induced system (Twomey *et al.*, 1978). On the other hand, the adherent system and PG-related system have both been found to be excessive in Hodgkin's disease (Goodwin and Messner, 1977).

It has also been suggested that a T-suppressor cell defect may be a primary immune deficiency in rheumatoid arthritis (Janossy *et al.*, 1981). A T-suppressor lymphocyte deficiency has been described in the rheumatoid synovium. These authors advance the theory that a T-suppressor cell defect may initiate a vicious cycle of interactions between inducer/helper T-cells and interdigitating macrophage-like cells. Such a situation could also include B-cell hyperactivation with production of immunoglobulins as well as the release of lymphokines and monokines causing tissue damage.

Whether or not a deficiency of the suppressor systems plays a part in the aetiology of

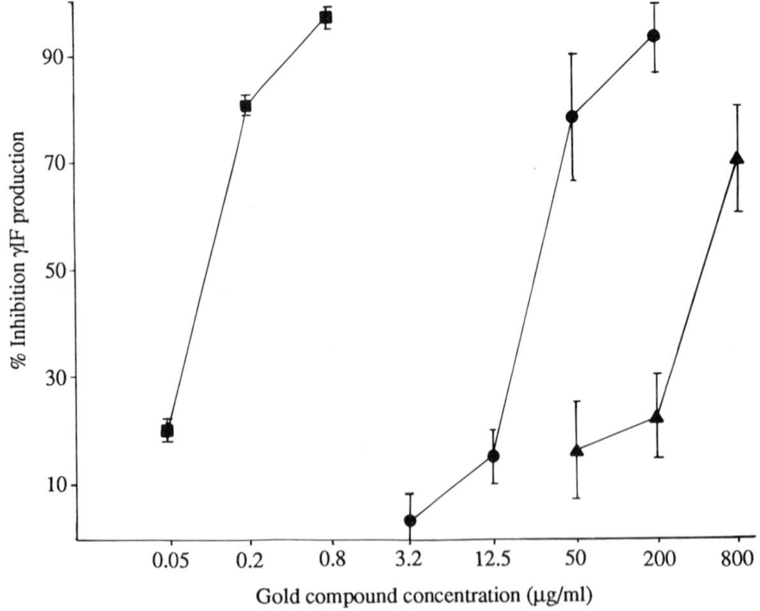

Fig. 2.19 Inhibition of sub-optimal PHA (0.1 μg/ml)-stimulated γ-interferon production by auranofin (■), sodium aurothiomalate (●) and aurothioglucose (▲). Each value is the mean ± SEM of 6 replicate cultures from two separate experiments

rheumatoid diseases, it seems clear that such a deficiency brought about by drug treatment might well lead to an exacerbation of T-lymphocyte-mediated injury and even immune complex formation. It is difficult to assess the clinical importance, and relation to other regulatory systems, of the PG-related T-suppressor system, and a possible direct PG-mediated inhibition of lymphocyte functions. However, there is enough circumstantial evidence to warn clinical rheumatologists of the dangers of treating patients chronically with non-steroidal anti-inflammatory drugs or cyclo-oxygenase inhibitors. There seems to be a strong possibility that they will exacerbate, perpetuate or even initiate a chronic arthritic condition.

While cyclo-oxygenase inhibitors enhance lymphocyte proliferation, the other classes of anti-rheumatic drugs tested in this investigation were all shown to exert powerful anti-proliferative activities at near therapeutic concentrations. The results are summarized in Table 2.1 and Fig. 2.20. Such effects have also been reported for anti-inflammatory steroids (Gordon and Nouri, 1981) although these agents interfere with PGE_2 production by monocytes/macrophages (Bray and Gordon, 1978). However, glucocorticoids also inhibit the secretion and activity of IL-2 and, in this respect, somewhat resemble PGE_2. Sub-optimal PHA-induced lymphocyte proliferation is selectively inhibited by both PGE_2 and the glucocorticoids, and selectively enhanced by both IL-2 and cyclo-oxygenase inhibitors.

In rheumatoid synovial cell populations, non-steroid anti-inflammatory drugs (NSAIDs) inhibit PGE_2 production at similar concentrations and with similar relative activities as seen in blood mononuclear cells. However, the release of collagenase by primary cultures of synovial cells is not inhibited at these therapeutic concentrations. In some such cultures, they were found to enhance collagenase release and the possibility arises that this effect is related to the presence of mononuclear cells in these heterogeneous populations in which endogenous PGE_2 formation mediates a negative-feedback influence. On the other hand, dexamethasone inhibited the synthesis of both PGE_2 and collagenase (Goodwin *et al.*, 1977). With few exceptions, the other anti-arthritic drugs tested showed no meaningful inhibition of PGE_2 production by synovial cells. However, by

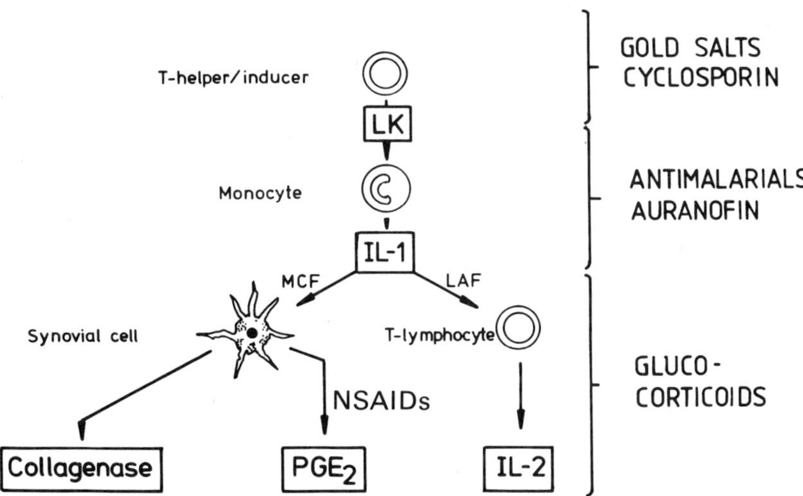

Fig. 2.20 Possible sites of action of anti-rheumatic drugs

Table 2.1 The effect of antimalarials, immunosuppressives and gold compounds on sub-optimal PHA-induced ^{3}H-TdR incorporation and IL-1 (MCF activity) release from human peripheral blood mononuclear cells, and on IL-1-induced release of PGE_2 from human rheumatoid adherent synovial cells

	IC_{50} (μM)		
Drug	*^{3}H-TdR Incorporation*	*MCF Release*	*PGE_2 Production*
Antimalarials			
Mepacrine	3.2	1.2	2
Chloroquine	13.0	1.8	>20
Immunosuppressives			
Cyclosporin-A	0.3	1.3	4
Azathioprine	3.0	15.0	>100
6-Mercaptopurine	2.0	20.0	>1000
Gold compounds			
Auranofin	1.4	0.9	3
Sodium aurothiomalate	350.0	100.0	~2000
Aurothioglucose	1000.0	700.0	>2000

suppressing the release of the monokine IL-1 some compounds interfered indirectly with inflammatory mediator production from synovial cells. In particular, the anti-malarials, mepacrine and chloroquine, were more active in inhibiting IL-1 production than lymphocyte proliferation. These effects were observed at therapeutically relevant concentrations and were accompanied by some inhibition of IL-1-induced PGE_2 synthesis by synovial cells, all of which could contribute to their anti-rheumatic activities. The more immunosuppressive agents, viz. cyclosporin-A, azathioprine and 6-mercaptopurine, were able to inhibit lymphocyte proliferation at lower concentrations than were effective against IL-1 release. Nevertheless, these drugs have been demonstrated to have anti-arthritic properties, whereas another anti-lymphoproliferative agent, cytosine arabinoside, which is not anti-arthritic, was unable to interfere with IL-1 production (Gordon *et al.*, 1982).

Cyclosporin-A, which is active only in immunologically-mediated models of inflammation (Borel *et al.*, 1978; Bolton and Cuzner, 1981), has been shown to suppress the release of macrophage-activating lymphokines from T-cells (Bunjes *et al.*, 1981). In the present studies some direct inhibitory effects on monocytes of high cyclosporin-A concentrations (10 μM) would be required to explain the level of inhibition of IL-1 release.

Sodium aurothiomalate (ATM) and aurothioglucose (ATG), the classical, injectable gold preparations, were only partially effective in inhibiting PHA-stimulated IL-1 release by mononuclear cells. We, therefore, examined their effects separately on spontaneous production by adherent monocytes/macrophages and PHA-induced release from non-adherent mononuclear cells, in which the lower level of spontaneous release of IL-1 by residual monocytes was increased approximately 3-fold by PHA-activated lymphocytes. The results indicated that ATM and ATG had little or no direct effect on monocytes, but selectively interfered with lymphocyte-mediated amplification of IL-1 release by monocytes. This was surprising since, on the basis of indirective evidence (Lipsky and Ziff, 1977), the anti-proliferative effects of ATM have been attributed to inhibition of the accessory role of monocytes/macrophages in lymphocyte mitogenesis, which are partly mediated via the monokine lymphocyte activating factor (LAF) (IL-1). Monocytes support lymphocyte proliferation not only through IL-1 production but also via cell–cell contact mechanisms which involve presentation of antigen and interaction of MHC class II antigen with responding lymphocytes. In considering the differences between the results of Lipsky and Ziff and the present author, it is possible that the effect which they observed, i.e. ATM inhibition of monocyte function, may be due to a mechanism involving MHC class II antigen, whereas the present study was looking specifically at another function, that of IL-1 production, which was found not to be inhibited by ATM. Nevertheless, the present results are consistent with the restricted anti-inflammatory properties of these agents, which are generally only effective in immunologically-mediated models of inflammation, e.g. adjuvant arthritis (Walz *et al.*, 1971) and experimental allergic encephalomyelitis (Gerber *et al.*, 1972).

However, the novel, orally-active, chrysotherapeutic agent auranofin (AF) inhibited spontaneous IL-1 release by monocytes although, as with the other gold compounds, there was slightly more activity against PHA-stimulated release. Not only was AF several hundred-fold more potent than the conventional gold salts, but also showed the exceptional property of inhibiting IL-1-induced PGE_2 synthesis by, and collagenase release from, synovial cells.

The findings with the gold salts have been confirmed and extended more recently (Gordon *et al.*, 1985; Barrett and Lewis, 1986) using a technique of replacement.

IL-2 was able to restore partly the anti-proliferative effects of all three gold compounds in their lower effective concentration range. However, this effect was limited, due to a direct action by the drugs upon IL-2 responder cells. Inhibition of IL-2 proliferation of PHA-blast cells was demonstrated in about the same concentration range as that which inhibited PHA-induced mononuclear cell proliferation. However, the addition of IL-1 failed to reverse the anti-proliferative effects of the gold salts, as would have been expected had inhibition of its release played a prominent role in this action of gold compounds.

The ability of γ-IF to restore low-dose anti-proliferative effects of ATM and ATG suggested a role for this lymphokine in the inhibitory effects of these drugs. Furthermore, ATM and ATG inhibited the release of γ-IF in these concentration ranges. AF also inhibited the release of γ-IF from mononuclear cell populations but the anti-proliferative effect of this drug was not affected by exogenous γ-IF. γ-IF has been reported to be produced by T-cells of the helper/inducer subset and to possess macrophage-activating properties (Martinez-Maza *et al.*, 1984; Sverdeshy *et al.*, 1984). Therefore, it appeared that, whereas all of the gold compounds blocked γ-IF production, AF additionally affected the ability of the macrophage to respond to the lymphokines.

Therefore, the gold salts tested displayed immunosuppressive activity *in vitro* which may represent their mode of action against rheumatoid arthritis. ATM and ATG inhibited lymphocyte function and lymphocyte-mediated amplification of macrophage function. AF inhibited both lymphocyte and macrophage function directly, and in addition inhibited production of mediators thought to be involved in rheumatoid joint destruction.

Summary

It is well established that cells of the lymphocyte and monocyte/macrophage series perform central roles in inflammation and the induction of immune responses. The biological activities attributed to cytokines imply that they could be important mediators of these functions. Inhibition of the extracellular secretion of cytokines by anti-arthritic compounds could therefore provide a mechanism whereby they exert their anti-inflammatory and immunomodulatory effects which would differ from those of cyclo-oxygenase inhibitors. For example, interference with the availability of IL-1 activities such as LAF, that promote lymphocyte function, would remove a positive stimulus that would balance the loss of the negative-feedback effect of endogenous PGE_2 production on lymphocyte activation. Furthermore, interference with IL-1 activities such as MCF would restrict inhibition of PGE_2 production to sites where lymphocyte-monocyte interaction was an important component. This might be advantageous in that undesirable side-effects attributable to non-specific inhibition of cyclo-oxygenase, e.g. gastro-intestinal irritation, could be avoided, but would be less effective where other inflammatory mechanisms triggered prostaglandin formation. However, interference with IL-1 release and/or action would reduce the levels of other synovial cell products apart from PGE_2 which contribute to the inflammatory process, e.g. collagenase. The agents shown here to affect IL-1 and IL-2 release include those that are regarded as capable of providing more than symptomatic relief of rheumatoid arthritis.

Non-steroid anti-inflammatory drugs (NSAIDs) and other anti-arthritic drugs have been compared with respect to their effects on human T-lymphocyte/monocyte/rheumatoid synovial cell interactions leading to inflammatory mediator production. NSAIDs inhibited prostaglandins (PGE_2) and enhanced sub-optimal phytohaemagglutinin (PHA)-stimulated tritiated thymidine (^{3}H-TdR) incorporation by mononuclear cells, although optimal responses were less affected. Exogenous interleukin-2 (IL-2) enhanced sub-optimal but not optimal PHA responses and the effects of the cyclo-oxygenase inhibitors were overcome by exogenous PGE_2. Thus, NSAIDs prevented the inhibition by endogenous monocyte-derived PGE_2 of IL-2 secretion and activity. Other anti-rheumatic drugs, including antimalarials, immunosuppressive agents and gold salts, inhibited PHA-induced lymphocyte proliferation regardless of the level of stimulation. Mepacrine and chloroquine were more effective in inhibiting the release of IL-1 that stimulated PGE_2 synthesis by synovial cells. Cyclosporin-A, azathioprine and 6-mercaptopurine were more potent as anti-proliferative agents than as inhibitors of mediator release. Sodium aurothiomalate and aurothioglucose selectively interfered with lymphocyte-mediated amplification of IL-1 release, whereas auranofin inhibited spontaneous production of monocytes and the action of IL-1 on synovial cells. In rheumatoid synovial cells, NSAIDs inhibited PGE_2 production but not collagenase release. Suppression of IL-1 release could lead indirectly to reduction of IL-2 and collagenase as well as PGE_2 production and consequently to more profound inhibition of immunologically-mediated inflammation.

References

Baker, P. E., Fahey, J. V. and Munck, A. (1981). Prostaglandin inhibition of T-cell proliferation is mediated at two levels. *Cell. Immunol.*, **61**, 52–61.

Barrett, M. L. and Lewis, G. P. (1986). Unique properties of auranofin as a potential anti-rheumatic drug. *Agents and Actions*, **19**(1/2), 109–15.

Bolton, C. and Cuzner, M. L. (1981). Modification of EAE by non-steroid anti-inflammatory drugs. In *The Suppression of Experimental Allergic Encephalomyelitis and Multiple Sclerosis*. Eds Davison, A. N. and Cuzner, M. L. pp. 189–97. London, Academic Press.

Borel, J. F., Wiesinger, D. and Gubler, H. U. (1978). Effects of the anti-lymphocytic agent cyclosporin A in chronic inflammation. *Eur. J. Rheum. Inflammation*, **1**, 237–41.

Bray, M. A. and Gordon, D. (1978). Prostaglandin production by macrophages and the effects of anti-inflammatory drugs. *Br. J. Pharmacol.*, **63**, 635–47.

Bray, M. A., Gordon, D. and Morley, J. (1978). Prostaglandins as regulators in cellular immunity. *Prostaglandins and Medicine*, **1**, 182–99.

Bunjes, D., Hardt, C., Solbach, W., Deusch, K., Rollinghof, M. and Wagner, H. (1981). Studies on the mechanism of action of cyclosporin A in the murine and human T-cell response *in vitro*. In *Cyclosporin A*. Ed. White, D. G. pp. 261–81. Amsterdam, Elsevier Biomedical.

Chouaib, S., Welte, K., Mertelsmann, R. and Dupont, B. (1985). Prostaglandin E_2 acts at two distinct pathways of T-lymphocyte activation: inhibition of interleukin-2 production and down-regulation of transferrin receptor expression. *J. Immunol.*, **135**(2), 1172–9.

Dayer, J-M., Krane, S. M., Russell, R. G. G. and Robinson, D. R. (1976). Production of collagenase and prostaglandins by isolated adherent rheumatoid synovial cells. *Proc. natn. Acad. Sci. USA*, **73**, 945–9.

Gerber, R. C., Whitehouse, M. R. and Orr, K. J. (1972). Effect of gold preparations on the development and passive transfer of experimental allergic encephalomyelitis in rats. *Proc. Soc. exp. Biol. Med.*, **40**, 1379–84.

Goodwin, J. S. and Messner, R. P. (1977). Prostaglandin-producing suppressor cells in Hodgkin's disease: an immunodepleting and immunosuppressive disorder. *N. Engl. J. Med.*, **297**, 963.

Goodwin, J. S., Bankhurst, A. D. and Messner, R. P. (1977). Suppression of human T-cell mitogenesis by prostaglandins. *J. exp. Med.*, **146**, 1719–34.

Gordon, D. and Nouri, A. M. E. (1981). Comparison of the inhibition by glucocorticosteroids and cyclosporin-A of mitogen-stimulated human lymphocyte proliferation. *Clin. exp. Immunol.*, **44**, 287–94.

Gordon, D. and Lewis, G. P. (1984). Effects of piroxicam on mononuclear cells. Comparison with other anti-arthritic drugs. *Inflammation*, **8** (Suppl.), S87–102.

Gordon, D., Bray, M. A. and Morley, J. (1976). Control of lymphokine secretion by prostaglandins. *Nature*, **262**, 401–2.

Gordon, D., Henderson, D. C. and Westwick, J. (1979). Effects of prostaglandins E_2 and I_2 on human lymphocyte transformation in the absence and presence of inhibitors of prostaglandin biosynthesis. *Br. J. Pharmacol.*, **67**, 17–22.

Gordon, D., Nouri, A. M. E. and Thomas, R. U. (1982). Effects of anti-arthritic drugs on the release of mononuclear cell factors (MCF) that stimulate prostaglandin (PG) E_2 production by rheumatoid synovial cells. *Int. J. Immunopharmacol.*, **4**, 311.

Gordon, D., Barrett, M. L. and Lewis, G. P. (1985). Effects of exogenous cytokines on the anti-lymphoproliferation effects of auranofin and Na aurothiomalate. *Br. J. Rheumatol.*, **24** (Suppl. 1), 224–5.

Janossy, G., Duke, O., Poulter, L. W., Panayi, G., Bofill, M. and Goldstein, G. (1981). Rheumatoid arthritis: a disease of T-lymphocyte/macrophage immunoregulation. *Lancet*, **ii**, 839–42.

Jose, P. J., Page, D. A., Wolstenholme, B. E., Williams, T. J. and Dumonde, D. C. (1981). Bradykinin-stimulated prostaglandin E_2 production by endothelial cells and its modulation by anti-inflammatory compounds. *Inflammation*, **5**, 363–78.

Lewis, G. P. (1983). Immunoregulatory activity of metabolites of arachidonic acid and their role in inflammation. *Br. Med. Bull.*, **39**, 243–8.

Lewis, G. P. and Barrett, M. L. (1986) Immunosuppressive actions of prostaglandins and the possible increase in chronic inflammation after cyclo-oxygenase inhibitors. *Agents and Actions* **19**, (1/2), 59–65.

Lipsky, P. and Ziff, M. (1977). Inhibition of antigen and mitogen induced lymphocyte proliferation by gold compounds. *J. clin. Invest.*, **58**, 455–66.

Martinez-Maza, O., Andersson, U., Andersson, J., Britton, S. and Deley, M. (1984). Spontaneous production of interferon-γ in adult and newborn humans. *J. Immunol.*, **132**(1), 251.

Rappaport, R. S. and Dodge, G. R. (1982). Prostaglandin E inhibits the production of human interleukin-2. *J. exp. Med.*, **155**, 943–8.

Rice, L., Laughter, A. H. and Twomey, J. J. (1978). Three suppressor systems in human blood that modulate lymphoproliferation. *J. Immunol.*, **122**(3), 991–6.

Shou, L., Schwartz, S. A. and Good, R. A. (1976). Suppressor cell activity after concanavalin A treatment of lymphocytes from normal donors. *J. exp. Med.*, **143**, 1100.

Sverdeshy, L. P., Benton, C. V., Berger, W. H., Rinderknecht, E., Harkins, R. N. and Palladino, M. A. (1984). Biological and antigenic similarities of murine interferon-γ and macrophage-activating factor. *J. exp. Med.*, **159**, 812.

Twomey, J. J., Laughter, A. H. and Steinberg, A. D. (1978). A serum inhibitor of immune regulation in patients with systemic lupus erythematosus (SLE). *J. clin. Invest.*, **62**, 173.

Walker, C., Kristensen, F., Bettens, F. and deWeck, A. L. (1983). Lymphokine regulation of activated (g_1) lymphocytes. *J. Immunol.*, **130**(4), 1770–3.

Walz, D. T., Di Martino, M. H. and Misher, A. (1971). The suppression of adjuvant-induced arthritis by gold sodium thiomalate in the rat. *Ann. Rheum. Dis.*, **30**, 303–16.

Webb, D. R. and Nowowiejski, I. (1981). Control of suppressor cell activation via endogenous prostaglandin synthesis: the role of T cells and macrophages. *Cell. Immunol.*, **63**, 321–8.

Webb, D. R., Rogers, T. J. and Nowowiejski, I. (1979). Endogenous prostaglandin synthesis and the control of lymphocyte function. *Ann. N.Y. Acad. Sci.*, **332**, 262–70.

CHAPTER 3

In vitro culture of human articular chondrocytes and intervertebral disc

M. T. Bayliss, J. McDowell, B. Johnstone and C. W. Archer

Introduction

Osteoarthritis and low back pain are undoubtedly the two most common reasons for incapacity to work. In 1982–83, low back pain alone accounted for 10% of the total work days lost and cost the UK National Health Service (NHS) £156 million. Considerable research effort has, therefore, been directed towards understanding the structure and metabolism of articular cartilage and intervertebral (I.V.) disc in health and disease. Most investigations have relied on studies *in vitro* and *in vivo* of animal tissues and these have provided a basic understanding of the biochemistry and physiology of cartilage and disc.

The articular cartilage and I.V. disc of laboratory animals, however, differ from those of humans in two very important ways: (i) they are usually immature and therefore the composition of the extracellular matrix is very different from adult humans, and (ii) the rate of turnover of the macromolecules in the matrix is very much faster than in adult humans. Both of these factors will influence strategies that may be developed for inducing repair of the tissue and investigations into the actions of therapeutic agents such as anti-rheumatic drugs.

Very few attempts have been made to use human cartilage as an alternative to animals, mainly because of difficulties in obtaining a regular supply of fresh tissue. Thanks to the cooperation of the authors' colleagues in the Department of Morbid Anatomy at the Institute of Orthopaedics, a very good supply of fresh human cartilage has been obtained. Organ-culture and cell-culture systems have been developed that maintain the zonal variations in metabolic activity and chemical composition that are present *in vivo*. Such systems could be used to gather information about drug action, or to study the biochemical events in early stages of disease, without resort to animal models.

Articular cartilage

Hyaline cartilage is a specialized connective tissue the major function of which depends on its state of hydration and the structural arrangement of a large extracellular matrix composed of collagen, proteoglycans, non-collagenous proteins and water. The biomechanical properties of normal articular cartilage, and its capacity to resist wear, result in part from the structure and properties of the fibrous composite formed when proteoglycans at high concentration are enmeshed with, and constrained by, a dense network of collagen fibres. Production and maintenance of the matrix depends on the cells within it, the chondrocytes, which make up only a small proportion (1–10%) of the tissue volume. The unique nature of the chondrocyte lies in its ability to form and regulate an organized tissue at sites remote from blood vessels, nerves and lymphatics.

Articular cartilage is not a homogeneous tissue. The composition of its organic matrix, the water content and the metabolic activity of the chondrocytes, all vary with depth from the articular surface to the subchondral bone. Furthermore, there are considerable variations in the composition and structure of the matrix in the pericellular and intercellular environments. The age and pathological state of the tissue also influence its composition.

CULTURE OF HUMAN ARTICULAR CHONDROCYTES

There have been numerous studies trying (a) to relate the properties of chondrocytes within articular cartilage to those maintained in culture, (b) to define the unique properties of chondrocytes, i.e. the repertoire of the articular cartilage phenotype as opposed to that of other cells, particularly fibroblasts, and (c) to develop culture conditions that accurately maintain the cartilage phenotype.

The major uses of cell culture are to study the synthesis and catabolism of matrix components and to determine their responsiveness to various cell factors, hormones and regulatory agents. Chondrocytes have been cultured as monolayers either on plastic or on modified substrates, in spinner culture and in agarose. These systems produce, at least for several passages, phenotypically stable chondrocytes when the source of cells is from immature rabbit, rat, porcine and bovine cartilage. However, the long-term stability of cultures of adult human chondrocytes has not been studied in detail, but changes in morphology to a fibroblastic appearance have been noted.

In an attempt to develop a culture system that maintains the chondrocyte phenotype and the biosynthetic activity of chondrocytes isolated from different layers of human cartilage, cells were cultured by three methods: (i) low-density monolayer, (ii) high-density spot culture micromass and (iii) suspension culture plated over agarose. The pattern of cell morphology and proteoglycan synthesis was distinct for each culture system.

Cultures were maintained for up to 3 weeks and were monitored regularly for changes in cell shape and their ability to synthesize large aggregating proteoglycans. In the examples shown in Fig. 3.1, chondrocytes were isolated from the femoral condyle cartilage of a 54-year-old male. Those plated in low-density cultures rapidly took on a fibroblastic appearance and by 10 days very few rounded cells remained. When the proteoglycans synthesized by these cells were examined, only a small proportion of them were capable of aggregating (Fig. 3.2). A similar effect on biosynthesis was observed for those cells in high-density micromass cultures; on day 10 of culture, mostly low-molecular-weight, non-aggregating proteoglycans, characteristic of fibroblastic cells were synthesized. The chondrocytes in high-density culture did, however, differ from the low-density cultures, in that they retained their polygonal appearance indicating that changes in phenotype are

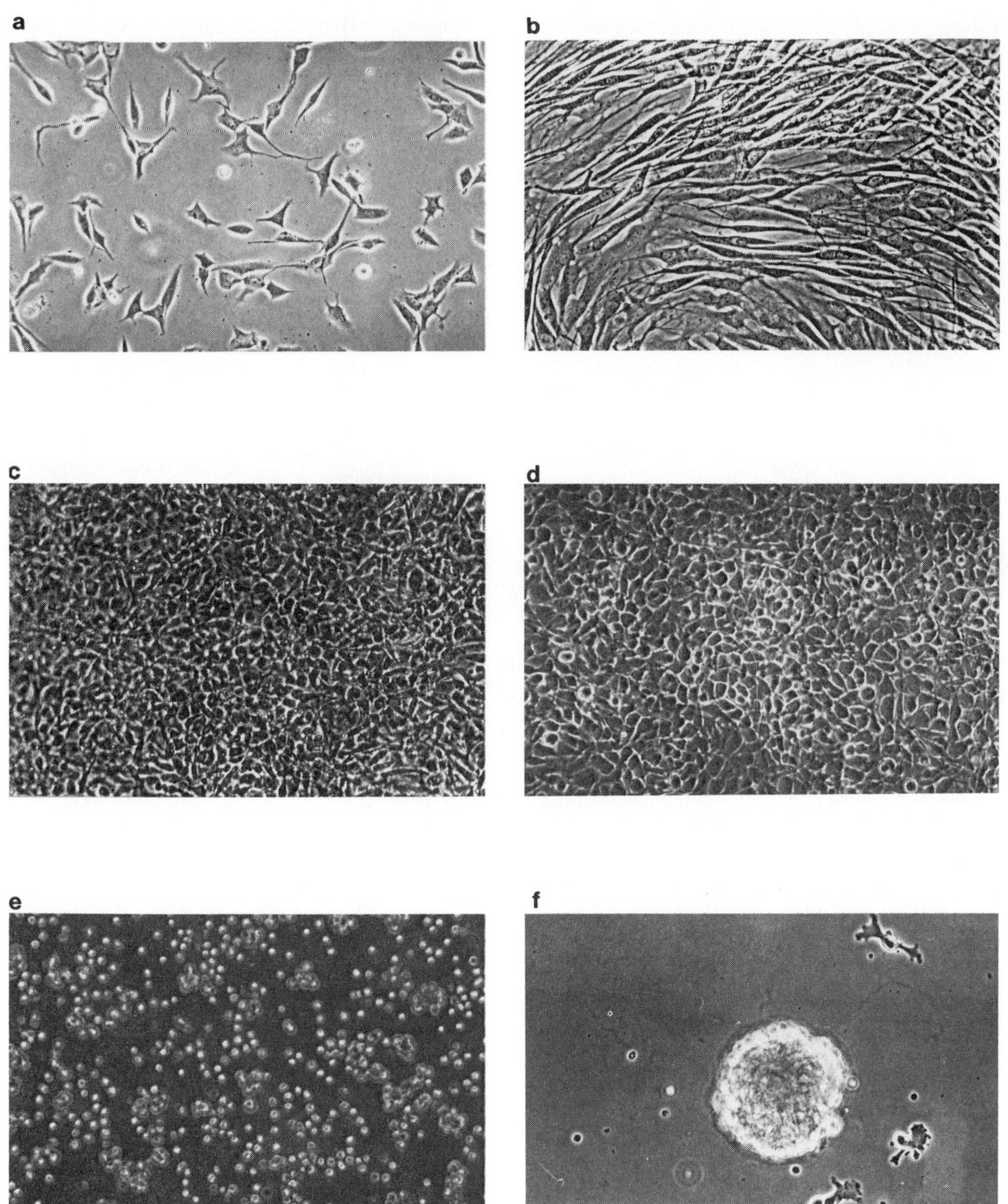

Fig. 3.1 Culture of human articular chondrocytes. (a) and (b), Low-density cultures at 3 days and 10 days. (c) and (d), High-density micromass at 3 days and 10 days. (e) and (f), Agarose cultures at 3 days and 10 days

not necessarily directly related to cell shape. The chondrocytes cultured over agarose retained their rounded appearance and formed large clusters of cells and continued to synthesize a cartilage-like matrix for up to 20 days.

The changes in biosynthetic activity observed in different culture systems have also been confirmed by analysing the matrix using specific monoclonal and polyclonal antibodies raised against substructures of the aggregating and non-aggregating proteoglycans. Only the agarose cultures continued to synthesize all matrix components. Thus, a culture method has been identified that can now be used to study the turnover of matrix macromolecules synthesized by human chondrocytes isolated from normal and diseased cartilage.

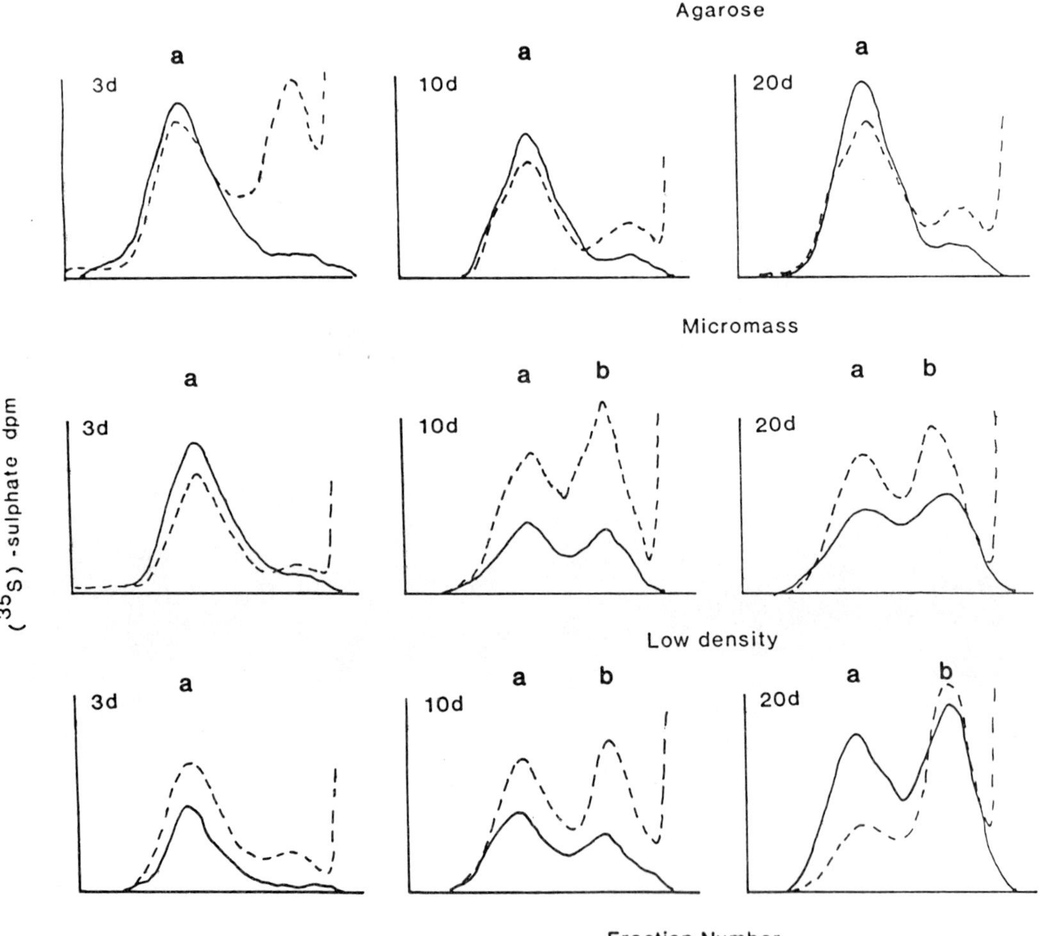

Fig. 3.2 Gel chromatography on Sepharose CL-2B of the medium (– – –) and cell-layer extracts (———) from cultures of human chondrocytes labelled with [^{35}S]-sulphate. Aggregating proteoglycans (a), non-aggregating proteoglycans (b)

Intervertebral disc

The integrity of the intervertebral disc is vital to the functioning of the spine. The discs serve both as joints and as shock absorbers, and account for about one-quarter to one-third of the length of the spinal column. The disc consists of two interconnecting tissues—the annulus fibrosus and the nucleus pulposus. In each case the extracellular matrix consists of an ordered arrangement of collagen, proteoglycan and non-collagenous proteins. The composition and structure of the proteoglycans vary between the tissues as does the organization of the collagen network. The nucleus has a very random arrangement of collagen fibres enmeshing the proteoglycan gel, whereas the collagen fibrils of the annulus are arranged into parallel bundles making up concentric lamellae around the nucleus. The lamellae of the outer annulus anchor directly into the bone and the direction of the collagen fibres alternates in neighbouring lamellae at an angle varying between 40° and 70° to the vertical axis. The collagen network provides the disc with the flexibility to act as a joint and yet enables it to withstand the high pressures induced during movement. The proteoglycans of the disc are similar to those of articular cartilage in that they form large aggregates. However, the age-related heterogeneity in the structure and composition of these molecules is more pronounced and occurs earlier in the disc.

Back pain is a common problem. It is responsible for over two million adults consulting their general practitioners each year in Britain alone (Editorial, *Lancet*, 1981) and yet in more than 70% of cases no firm diagnosis can be made (Wells, 1983). The treatment offered to relieve back pain is therefore largely empirical. Progress in diagnosis and management of spinal disorders will not be possible until we gain further basic knowledge of the function and behaviour of spinal components both in health and disease.

It is clear that abnormalities of the intervertebral disc are implicated in some common, but nevertheless important, back problems, such as disc herniation and disc instability and may have a significant role in others like idiopathic low back pain. The functional properties of the intervertebral disc, like that of articular cartilage, depend on its extracellular matrix. Proteoglycans, through their fixed negative charge, imbibe water and exert a high swelling pressure keeping the collagen network extended. Furthermore, because they are present in high concentrations, they restrict the rate of fluid loss and modulate the diffusion of larger solutes. Thus, knowledge of the structure of human intervertebral disc proteoglycans is a prerequisite for the understanding of the mechanical functions of the normal disc. Although some limited progress has been made in understanding the changes in proteoglycan chemistry due to age, disease and distribution within the human disc, there is virtually no information on the synthesis of proteoglycans. For proteoglycan concentration to remain constant, the rate of synthesis must balance that of degradation. However, the proteoglycan content of the human disc decreases markedly with age and also in degenerative conditions such as spondylolisthesis (Roberts *et al.*, 1982). It is not known if this change in composition results from an increase in the rate of proteoglycan degradation or a decrease in the rate of synthesis. Proteoglycan synthesis rates have been calculated from the results of tracer tests *in vivo* for some animal species (guinea pig, rabbit, dog, mouse) (Davidson and Small, 1963; Lohmander *et al.*, 1973; Urban *et al.*, 1978; Venn and Mason, 1983), but more recently a reliable *in vitro* method has been developed using a technique which controls disc hydration and prevents proteoglycan loss, factors which could influence cell metabolism (Bayloss *et al.*, 1986). Using this method synthesis rates *in vitro* for rabbit and dog discs were shown to be similar to those measured *in vivo*. As a result of these studies, it became clear that only by carefully controlling the swelling of the tissue during culture are meaningful rates of synthesis obtained. We have also used this new culture method to measure the synthesis rates in human intervertebral disc, and to show how features such as hydration, age, pathology and position within the disc affect the biosynthesis of proteoglycans (Bayliss *et al.*, 1988).

THEORETICAL BACKGROUND

In vivo the intervertebral disc is always under an external load that changes with posture. Because fluid can interchange between the disc and the surrounding blood vessels, the fluid content of the disc, P_s, balances the load, P_a:

$$P_a = P_s$$

P_s arises from the difference between the osmotic pressure of the proteoglycans, which imbibe water and hence attempt to swell the tissue, and the tensile forces of the collagen network, which resist swelling. The magnitude of P_s at post-mortem is about 0.1–0.2 MN/m^2 (1–2 atm.), which is equivalent to the external load on the lumbar discs in the prone position (Nachemson and Elfström, 1970; Urban and McMullin, 1985). When the disc is excised and is placed in aqueous solution, e.g., physiological saline, P_a becomes zero. Consequently, the disc swells in order to adjust P_s to zero. As the fluid content increases, the proteoglycan concentration falls and hence the osmotic pressure decreases. At the same time, as the volume of the disc increases, the tension in the collagen network increases. Swelling continues until the reduced osmotic pressure is balanced by the increased collagen network tension. An additional factor in the disc is that proteoglycans leach from the swollen tissue, and hence proteoglycan loss follows swelling. This lessens the osmotic pressure. Disc slices *in vitro* can double their volume and lose more than 50% of their proteoglycans. If the disc slices are placed in a solution of high osmotic pressure (Π mm) instead of into normal saline, the fluid content will respond in the same way as to a mechanical load, i.e. until

$$P_s = \Pi \text{ mm}$$

If Π mm is greater than P_s, the disc will lose fluid, and if Π mm is less than P_s of the post-mortem disc slice, the tissue will swell. Thus, by finding an iso-osmotic bathing solution the tissue will retain its post-mortem hydration. High-molecular-weight dextrans, which are sterically excluded from such tissues, have been used to maintain iso-osmotic conditions in cornea and muscle (Dohlman and Anseth, 1969). However, at the osmotic pressures of several atmospheres necessary to balance the swelling pressures of load bearing cartilages, dextran solutions are too viscous to handle. The osmotic properties of polyethylene glycol 20 000 (PEG) have been well characterized (Edmond and Ogston, 1968), and concentrated solutions are less viscous than those of dextran. Although PEG can penetrate into cartilages, it can be excluded by interposing a semi-permeable membrane such as small-pore dialysis tubing between the tissue and the PEG solutions (Urban and Maroudas, 1979). Hence, PEG has been used in this study, and the incubation of all disc samples was carried out with the tissue enclosed in dialysis sacs.

IN VITRO CULTURE OF HUMAN INTERVERTEBRAL DISC

The amount of swelling and proteoglycan released from slices of anterior annulus incubated in 0.15 M NaCl is considerable. After only 10 min in solution the slices had swollen by 50% and by 30 min they had doubled their wet weight. Proteoglycan loss also increased with time; after incubation, 25% of the tissue glucuronic acid was released into the medium.

The hydration changes in slices of anterior annulus, enclosed in dialysis sacs, were determined after incubating them in solutions containing different concentrations of PEG (Fig. 3.3). The higher the PEG concentration in the medium the less the annulus swelled, until a point was reached where the osmotic swelling pressure of the tissue was balanced by that of the PEG solution and the hydration of the slice was maintained at its *in vivo* level. Polyethylene glycol concentrations higher than this caused the hydration to fall below the *in vivo* level as the osmotic imbalance between the medium and the tissue increased. The inherent variability of human specimens, even those of the same age, meant that it was not possible to predict the exact composition, and consequently the swelling pressure of each specimen, before culturing it. This was even more apparent for pathological discs where it was impossible to predict the matrix composition. The PEG concentrations required to control tissue hydration were therefore determined experimentally for each disc at the time of culture. Thus, several slices of annulus or nucleus from each disc studied were cultured in [^{35}S]-sulphate labelled medium containing different concentrations of PEG and only slices which had a hydration, at the end of incubation, within $\pm 10\%$ of the *in vivo* value, were used to calculate the rate of proteoglycan synthesis. Previous work has shown that hydration changes of less than $\pm 10\%$ have a minimal effect on the rate of sulphate uptake by rabbit and dog disc (Bayliss *et al.*, 1986).

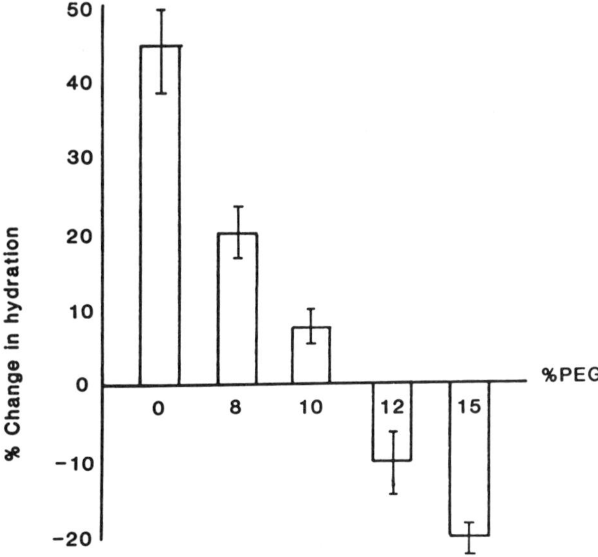

Fig. 3.3 Change in disc hydration with increased polyethylene glycol (PEG) concentration. Slices of adult human anterior annulus were incubated for 4 h in 0.15 M NaCl containing different concentrations of PEG

NORMAL (AGE-RELATED) AND PATHOLOGICAL CHANGES IN PROTEOGLYCAN SYNTHESIS RATES

Slices of normal adult annulus, cultured under the conditions described above, incorporated [^{35}S]-sulphate at a linear rate for up to 4 h (Fig. 3.4). The rates of proteoglycan synthesis for slices of normal annulus from discs of different ages collected at 18–24 h post-mortem are shown in Table 3.1. Foetal and newborn nucleus had rates that were 2–3 times higher than the annulus from the same discs, whereas synthesis rates for the adult annulus were 5 times less than those of immature annulus. The rate of [^{35}S]-sulphate incorporation into slices of annulus from pathological discs was also determined. Spondylolisthetic discs taken from level L_5–S_1 had lower rates than those from level L_4–L_5; this difference was not related to the severity of the lesion. No differences in rates related to disc level were evident for the other pathologies. Within all pathological groups there was a wide range of biosynthetic activity. Many of the pathological specimens had higher rates of synthesis than post-mortem discs in the same age range, however, this may not be related to the disease state but may simply reflect the different time that had elapsed prior to incubation.

The rationale for measuring synthesis rates in pathological discs is based on the assumption that cells in the diseased tissue will attempt to repair the damaged matrix. In this respect the strategy is analogous to that proposed for chondrocytes in degenerative joint disease and suffers from some of the same pitfalls. This is illustrated by the large scatter

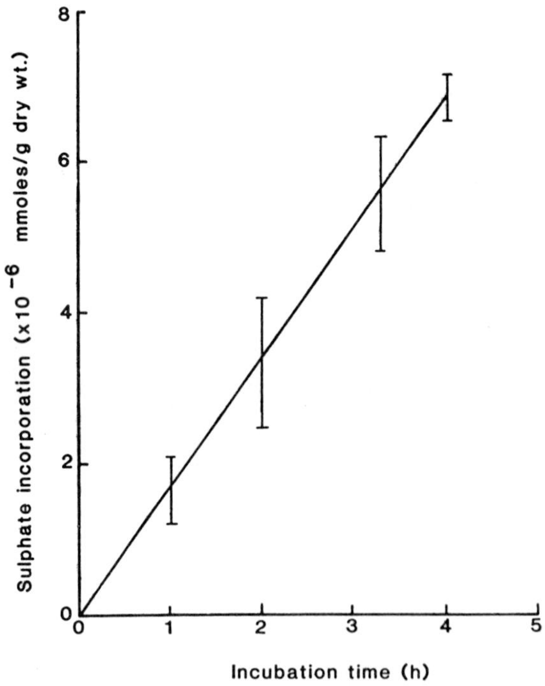

Fig. 3.4 Linearity of [^{35}S]-sulphate incorporation with time in culture. Slices of adult human anterior annulus were incubated for different time periods in [^{35}S]-sulphate-containing medium. All values for incorporation are for slices where hydration was retained within $\pm$10% of a non-incubated control

Table 3.1 Age-related changes in the rate of [^{35}S]-sulphate incorporation by human post-mortem discs

Age	*Rate of synthesis* ($\times 10^{-6}$ mmoles/h/g dry wt)	
	Annulus	*Nucleus*
Foetal	15.6	40.5
6 weeks	17.2	40.8
32 years	3.1	–
50 years	2.2	–

in the rates measured, which reflects not only the heterogeneity of the disease processes, but also the problems encountered in defining the biological stage of disc degeneration.

TOPOGRAPHICAL VARIATIONS IN THE RATE OF PROTEOGLYCAN SYNTHESIS

In order to determine the rates of proteoglycan synthesis at an *in vivo* hydration level for all regions across the annulus, serial slices were incubated in different PEG concentrations and were then sectioned from outer to inner annulus after incubation. The post-incubation hydration of each sagittal section was then determined and these values were compared with the hydration profile of a non-incubated slice. The slice with a hydration level after incubation closest to that of the control slice was used to calculate the rate of proteoglycan synthesis. Overall, the profiles indicate that there is a maximum rate of proteoglycan synthesis in the mid-annulus region of the adult human disc and that this is most clearly seen in the slices with hydration profiles closest to that of the control (Fig. 3.5). Confirmation of the topographical variation in synthesis rates was obtained by autoradiography of [^{35}S]-sulphate-labelled tissue slices, and indicates that the cells in the central region of the annulus contribute most to the proteoglycan content of the adult disc.

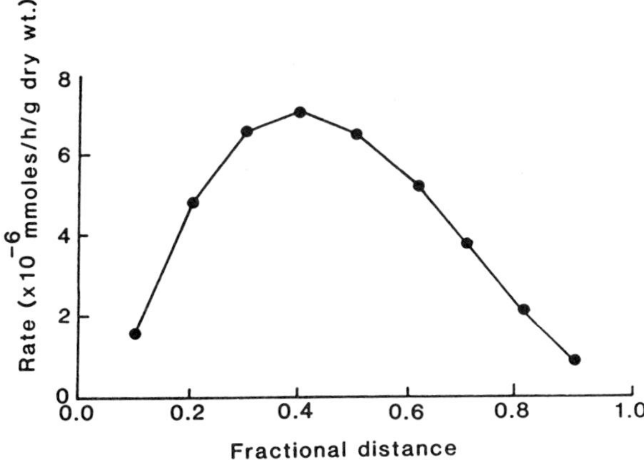

Fig. 3.5 Variation in the rate of proteoglycan synthesis at *in vivo* hydration, across the anterior annulus of a 32-year-old post-mortem human disc

Summary

The results presented in this paper give, for the first time, values for the rates of proteoglycan synthesis in human intervertebral disc. They also illustrate the difficulties that are encountered in any study of the biological properties of such a heterogeneous tissue. It is very important that any investigation of the biochemistry of the human disc should take account of the variations in composition, cellular activity and tissue organization and the way in which these parameters change with age, region and disc level.

References

Bayliss, M. T., Urban, J. P. G., Johnstone, B. and Holm, S. (1986). *In vitro* method for measuring synthesis rates in the intervertebral disc. *J. Orthop. Res.*, **4**, 10–17.

Bayliss, M. T., Johnstone, B. and O'Brien, J. P. (1988). Proteoglycan synthesis in the human intervertebral disc. Variation with age, region and pathology. *Spine* (in press).

Davidson, E. A. and Small, W. (1963). Metabolism *in vivo* of connective tissue mucopolysaccharides. Chondroitin sulphate C and keratosulphate of nucleus pulposus. *Biochim. Biophys.* Acta, **69**, 445–52.

Dohlman, C. H. and Anseth, A. (1969). The swelling pressure of ox corneal stroma. *Acta Ophthal. Kbn.*, **35**, 73–84.

Editorial (1981). Progress in back pain? *Lancet*, **i**, 977–9.

Edmond, E. S. and Ogston, A. G. (1968). Phase separation in ternary aqueous systems. *Biochem. J.*, **109**, 569–76.

Lohmander, L. S., Antonopoulos, C. A. and Friberg, U. (1973). Chemical and metabolic heterogeneity of chondroitin sulphate and keratan sulphate in guinea pig cartilage and nucleus pulposus. *Biochim. Biophys. Acta*, **304**, 430–48.

Nachemson, A. and Elfström, G. (1970). Intravital dynamic pressure measurements in lumbar discs. *Scand. J. Rehab. Med.*, Suppl. 2, 50–83.

Roberts, S., Beard, H. K. and O'Brien, J. P. (1982). Biochemical changes in intervertebral discs in patients with spondylolisthesis and with tears of the annulus fibrosus. *Ann. Rheum. Dis.*, **44**, 78–85.

Urban, J. P. G. and Maroudas, A. (1979). Measurement of fixed charge density and partition coefficients in the intervertebral disc. *Biochim. Biophys. Act*, **586**, 166–78.

Urban, J. P. G. and McMullin, J. (1985). Swelling pressure of the intervertebral disc. *Biorheology*, **22**, 145–57.

Urban, J. P. G., Holm, S. and Maroudas, A. (1978). Diffusion of small solutes into the intervertebral disc: an *in vivo* study. *Biorheology*, **15**, 203–23.

Venn, G. and Mason, R. M. (1983). Biosynthesis and metabolism *in vivo* of intervertebral disc proteoglycans in the mouse. *Biochem. J.*, **215**, 217–25.

Wells, N. (1983). *Back Pain.* Whitehall, London, UK, Office of Health Economics.

CHAPTER 4

Progress in the study of uptake and transport of exogenous material by nerve cells in tissue culture

J. N. Payne

Neurones, in both the peripheral and central nervous systems, have the ability to transport materials along their axons. This 'axonal transport' is bidirectional so that material may be moved either away from or towards the cell body. The phenomenon is of fundamental biological importance in view of the enormous relative distance of the cell body from the axon terminal (Fig. 4.1).

Thus, for example, the process of protein synthesis in the neurone occurs exclusively in the cell body. All structural protein and protein for enzymes and neurotransmitters has to be manufactured in the cell body and transported to its final destination. Similarly, the neurone is dependent upon retrograde axonal transport in order to acquire information about the functional state of its axon and axon terminal. Retrograde axonal transport is particularly important in development because it signals to the cell body of the neurone that the axonal terminal has correctly arrived at its target. Trophic substances, such as nerve growth factor, are believed to be involved in this process. Failure of growing axons to reach their target tissue and to transport retrogradely nerve growth factor, results in a reduction of cell number and in enzyme activity in sympathetic neurones (Hendry and Campbell, 1976). A similar mechanism may act in the initiation of the chromatolytic response in the neuronal cell body to axonal injury.

This chapter is concerned with progress in a project designed to provide further information about retrograde axonal transport. One of the first pieces of experimental evidence for such a phenomenon was that large protein molecules, such as albumin and horseradish peroxidase, were actively transported from the terminals of neurones to their cell bodies (Kristensson 1970a; Kristensson and Olsson, 1971). It also became clear that even larger biological units such as viruses were taken up and transported in this way (Kristensson, 1970b). Similarly, molecules of clear relevance to neurones, such as nerve growth factor, are retrogradely transported (Thoenen and Schwab, 1978). What came as a surprise, however, was that there is a large range of relatively small molecules which are retrogradely transported by neurones. Kuypers *et al.* (1977) screened a range of fluorescent compounds and showed that many, such as Evans Blue, propidium iodide, primulin and

DAP I, were taken up and transported to the cell body after the axon terminals were exposed to them.

This immediately raises the question of what sort of chemical substances are dealt with in this way by the neurone. In other words, what physico-chemical properties do these small molecules have to possess in order that they should be retrogradely transported? In attempting to answer such a question it is useful to adopt a similar strategy to the structure–activity studies of pharmacologists. Apart from the basic biological importance of this question there are other, more clinically orientated, objectives involved in understanding what sort of molecules are retrogradely transported. Some substances, for example, ricin and doxorubicin (Koda and Van der Kooy, 1983) exert toxic effects because they are so transported, and there is a real risk that some drugs and other small molecules such as food dyes and additives could have a similar effect. In this context, it is useful to bear in mind that other molecules, such as the fluorescent tracer true blue, can remain unaltered in neurones for many months after having reached their cell bodies by retrograde axonal transport (Innocenti, 1981).

It has been suggested that it would be useful selectively to introduce anti-viral drugs into neurones in recurrent infections with Herpes and similar viruses, which remain resistant to

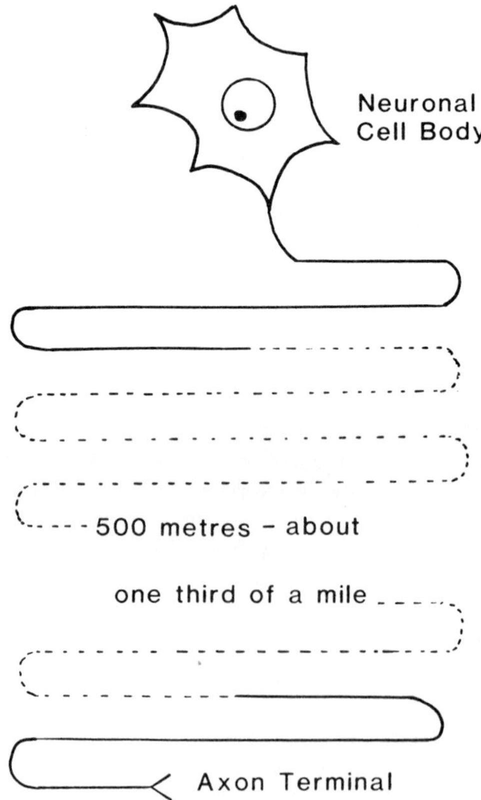

Fig. 4.1 A motor neurone with the length of its axon at the same scale as the size of its cell body. Axonal transport is clearly essential to allow communication between the cell body and its far distant axon terminal

treatment by being sequestered in neuronal cell bodies (Haschke *et al.*, 1980). Clearly to do so requires knowledge of what sorts of molecules are so transported. Finally, the results will be important in the understanding of the aetiology of peripheral neuropathies in which retrograde axonal transport is affected (Jakobsen *et al.*, 1986).

To carry out such an investigation, the substances which are being investigated must be rendered visible. Fortunately, a wide enough range of compounds are fluorescent and can easily be demonstrated and quantified by fluorescence microscopy. Thus, such fluorochromes are detectable in dilutions a thousand times greater than that of simple coloured dyes (Hamperl and Schümmelfeder, 1961). Accordingly, the uptake and transport of a large number of fluorescent compounds is being investigated.

The question which is particularly relevant here, however, is whether these uptake and transport phenomena can be investigated in ways which reduce, or better still, eliminate the need to use experimental animals. Most of the work on axonal transport up until now has used whole animals. Thus, Kristensson (1970a) investigated retrograde transport of labelled albumin and horseradish peroxidase (Kristensson and Olsson, 1971) by injecting small volumes of a solution of these substances into the tongue of young rodents. After a few days the animals were killed and it was found that there were labelled neurones in the hypoglossal nucleus showing that these substances had been retrogradely transported. Other investigations on axonal transport used a similar experimental protocol: that is injection of a tracer substance; allowing a suitable survival period; and then killing the animal to determine to what extent the tracer had been transported by active processes within the relevant neurones. Other common investigative approaches are to ligate a whole nerve in a live animal and observe the build-up of transported materials at the ligature site; or to use special microscopical techniques to observe particulate transport within individual axons of living animals.

Experimental techniques which do not use whole animals have mainly concentrated on the microscopic observation of axonal transport in neurites of cells in tissue culture. This method can be used successfully to study the dynamics of organelle movement in axons. In particular, the velocity and saltatory nature of the movement of particles has been investigated in this way (reviewed by Forman, 1982). Other investigations *in vitro* of axonal transport have used whole nerve preparations from animals (Ochs, 1972). These, while being more satisfactory than operating and experimenting on the animals whilst they are still alive, do require that animals be killed to obtain the nerve preparations.

The aim here will be to discuss some methods which the present author has been developing, to look at these processes using not whole animals but cells in culture. It seems most relevant to concentrate on the methods rather than the results at this stage.

Before substances can be retrogradely transported they have first to get into the cell. The extent to which a substance binds to the cell membrane and the nature of its molecular charge both seem to be important in determining whether or not uptake and transport will occur. It seemed appropriate, therefore, to develop a system for investigating cellular uptake of fluorescent substances which are known to be retrogradely transported. Initially, it was decided to study how these compounds become internalized in a simple cell model, the chick fibroblast. Using 8-day-old fertilized hen's eggs a fibroblast cell line was set up and maintained in cell culture. This enabled large numbers of cells to be obtained without using any further live eggs. The Humane Research Trust enabled the purchase of a laminar flow chamber and an incubator to allow this cell culture work.

The cells were grown in standard tissue culture flasks, but for the actual experimental procedures they were allowed to adhere to, and multiply on, the surface of glass coverslips

(Fig. 4.2) so that they were present in large numbers but were not yet confluent. Thus, although there were many cells present, many were completely separate from their neighbours.

The cell chamber illustrated in Fig. 4.3 and 4.4 was constructed from 3- and 6-mm thick perspex sheet. The central well was maintained at 37°C by circulating water from a thermostatically-controlled water-bath through the channels surrounding it. A small electronic temperature probe was used to check the temperature within the central well. A coverslip, onto which cells had adhered and grown, was sealed over the central well using silicone vacuum grease (Dow-Corning). Two inlet tubes led to the chamber created which had a volume of 0.25 ml. One such tube was connected by fine plastic tubing to a universal container filled with phosphate-buffered saline solution, while the other was connected to a similar reservoir of a solution of the fluorochrome under investigation. Both reservoirs were kept at 37°C by heated water-bath. The outlet tube from the chamber was connected, by similar plastic tubing, to a 50-ml syringe operated by a push–pull syringe pump (Harvard Instruments).

A Leitz Orthoplan fluorescence microscope, fitted with a high-pressure mercury light source (HB 100), was used to examine the cells on the underside of the coverslip. The appropriate mirror-filter system (Ploempak) for incident illumination of the fluorochrome

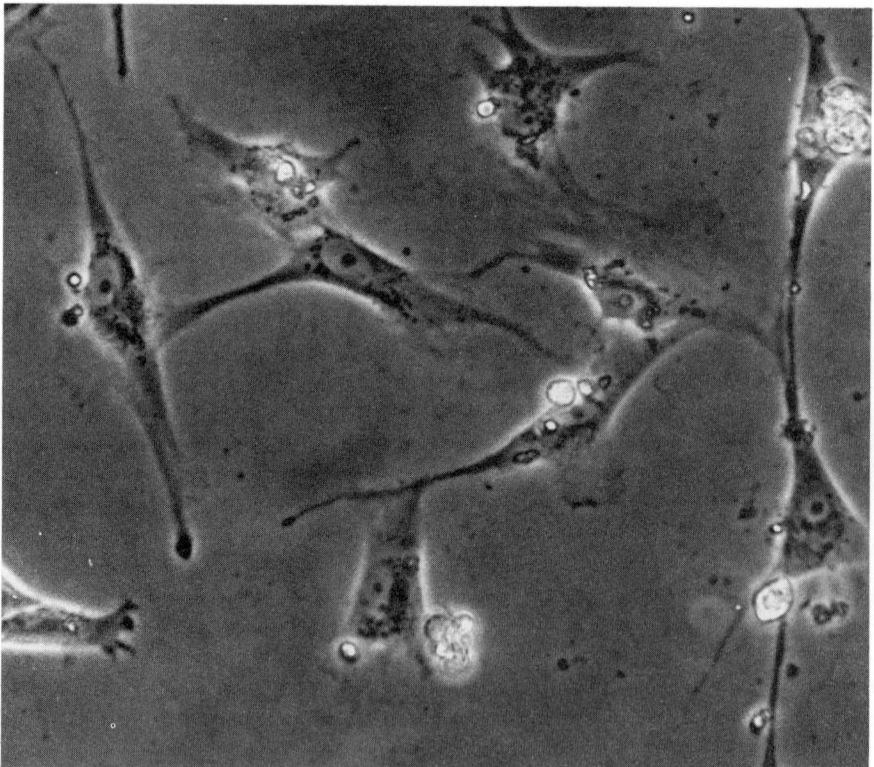

Fig. 4.2 A phase-contrast micrograph of fibroblasts grown on a coverslip for investigation of fluorochrome uptake in the cell chamber described. (×400)

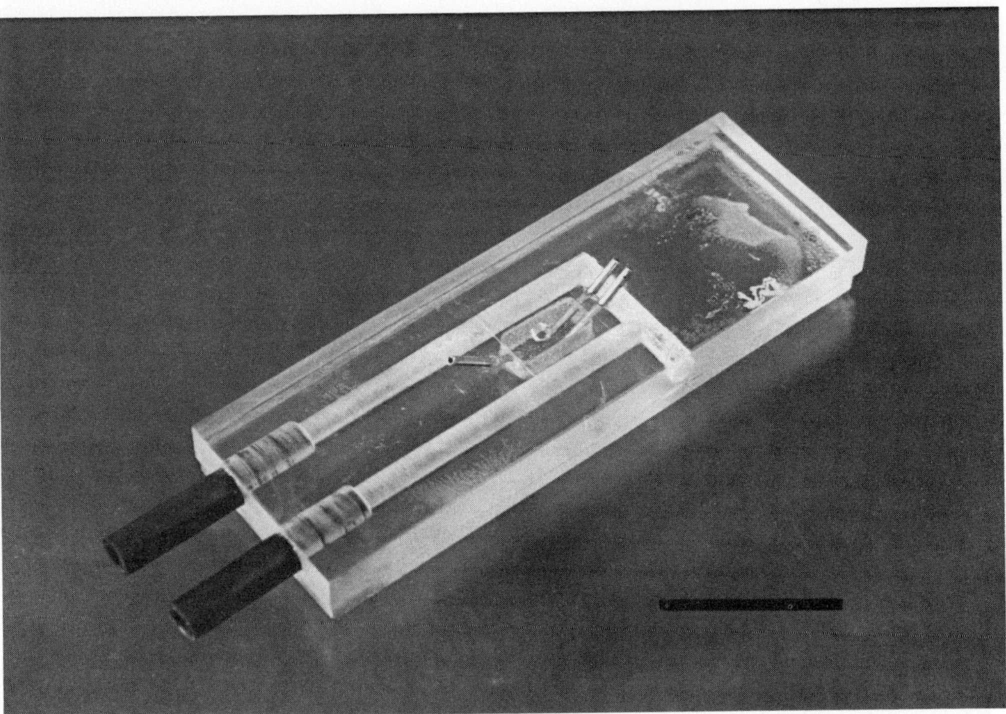

Fig. 4.3 The cell chamber used for investigation of uptake of fluorescent compounds in living cells. The bar is 20-mm long

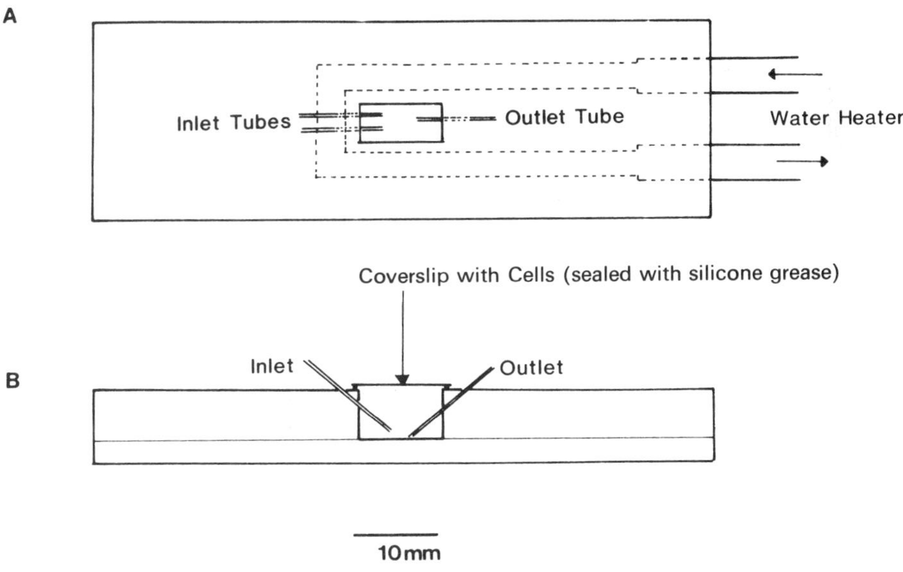

Fig. 4.4 A scale drawing of the cell chamber seen (A) from above and (B) from the side

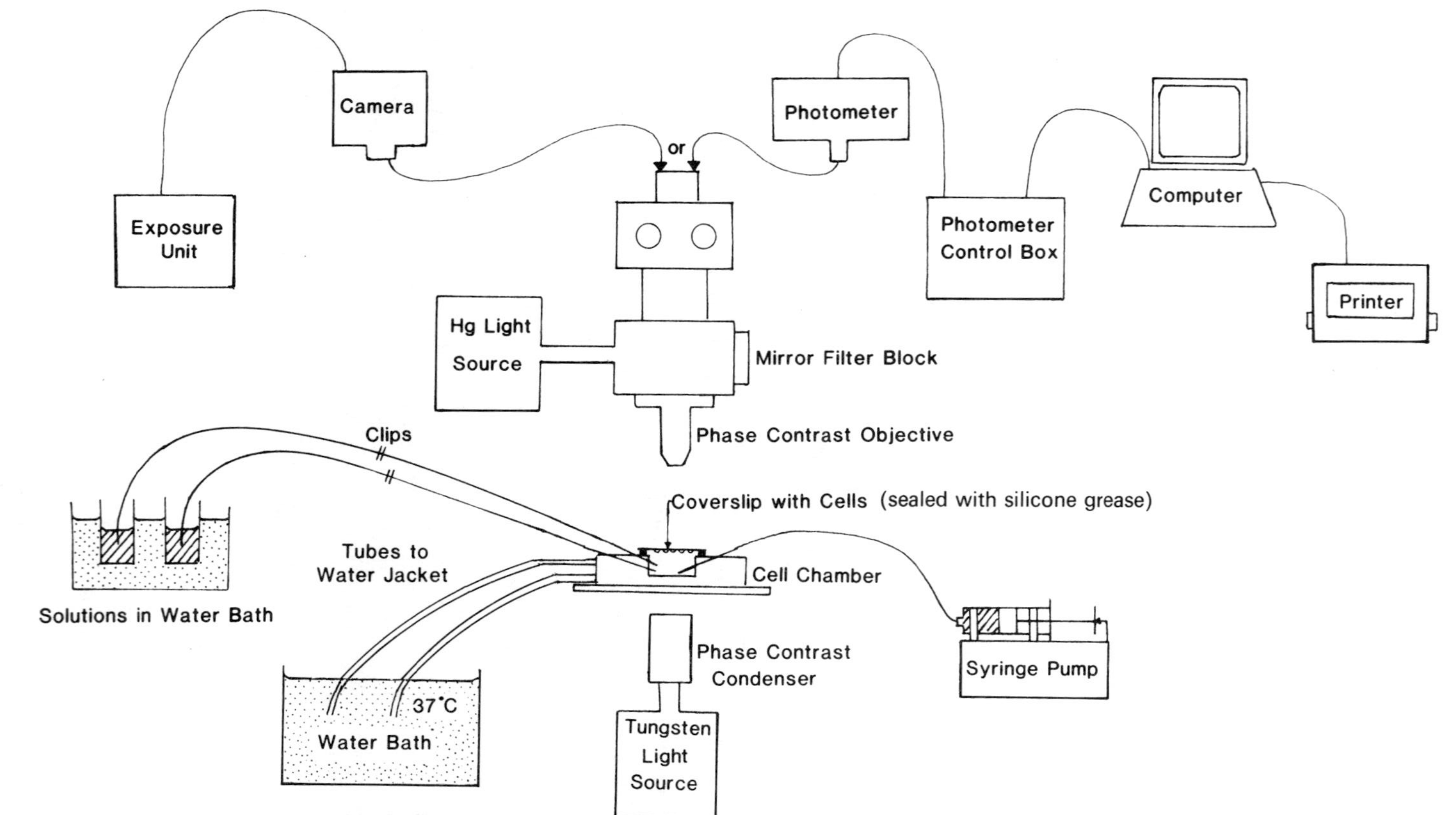

Fig. 4.5 The whole experimental system shown diagrammatically

under investigation was selected; in the case of Evans Blue, one providing excitation with green light (530–560 nm, Leitz mirror-filter block N2) was used. A shutter in the excitation light pathway ensured that the cells were only exposed to this light when photographs or photometric readings were made. Phase-contrast illumination with white light was also available, and the chamber was designed to ensure that it was still possible to align the phase rings in the condenser and objective, and so obtain a clear phase-contrast image of the cells on the underside of the coverslip. The microscope had a trinocular head, so either a camera or a photometer could be attached.

Photographic recording of the uptake and fate of fluorochrome under investigation involved taking micrographs at fixed time intervals throughout the period of exposure to the fluorochrome. All micrographs in a given time sequence were taken using the same exposure time.

A Leitz MPV Compact II photometer was linked to a Zenith Z100 microcomputer for which a program was written to take photometric readings automatically at fixed time intervals for a given length of time. These data were stored on floppy disc as well as being printed. For photometric measurements an aperture was selected which gave a circular measuring window, 6 μm in diameter, when a 40× objective was used. The field diaphragm was closed to provide an illuminated field as close in size as possible to that of the measuring aperture. The sensitivity of the photometer was adjusted both before and between each measurement session by calibrating against a piece of fluorescent uranyl glass (Ploem, 1970). The whole experimental system is illustrated diagrammatically in Fig. 4.5.

To investigate the uptake and fate of a particular fluorochrome the following procedure was adopted. The well of the cell chamber was filled with phosphate-buffered saline solution, and a coverslip with attached cells was sealed over the opening with silicone grease. Using phase-contrast illumination, the cell layer was identified and brought into sharp focus. The microscope stage was moved so that a suitable group of cells was in the camera field, or a single cell was positioned under the measuring window of the photometer system. Next, about 2 ml (which represents eight times the chamber volume) of the solution of fluorochrome was drawn into the chamber by slowly (4 ml/min) withdrawing the syringe connected to the outlet tube. In this way the cells could be examined both before and from the very moment at which they became exposed to the fluorochrome. A sequence of photographs or photometric readings was begun when the fluorochrome was introduced into the cell chamber so that the time course of accumulation of the substance into the cell could be determined. In some experiments the fluorochrome was then replaced once more by phosphate-buffered saline solution to investigate possible migration of the fluorochrome from the cell.

The apparatus proved easy to use and the chamber was found to be held at the correct temperature. Fibroblasts could be clearly visualized using phase-contrast before being exposed to the fluorochrome under investigation. A typical series of photographs, in this case illustrating uptake of 0.01% Evans Blue from phosphate-buffered saline solution, is shown in Fig. 4.6. There is initially no fluorescence in the cell, but after about 40 sec the cell surface becomes labelled. Later still the cytoplasm, but not the nucleus, has acquired the fluorochrome; and finally fluorescence becomes localized to cytoplasmic granules.

In Fig. 4.7 some results of measuring the intensity of fluorescence in the cytoplasm after exposure to the same fluorochrome are plotted over a period of 20 min. A very short lag period is followed by a period of approximately linear increase and finally there is a prolonged plateau during which the fluorescence remains unchanged.

It is not intended to discuss the results presented here in any detail but it is clear that

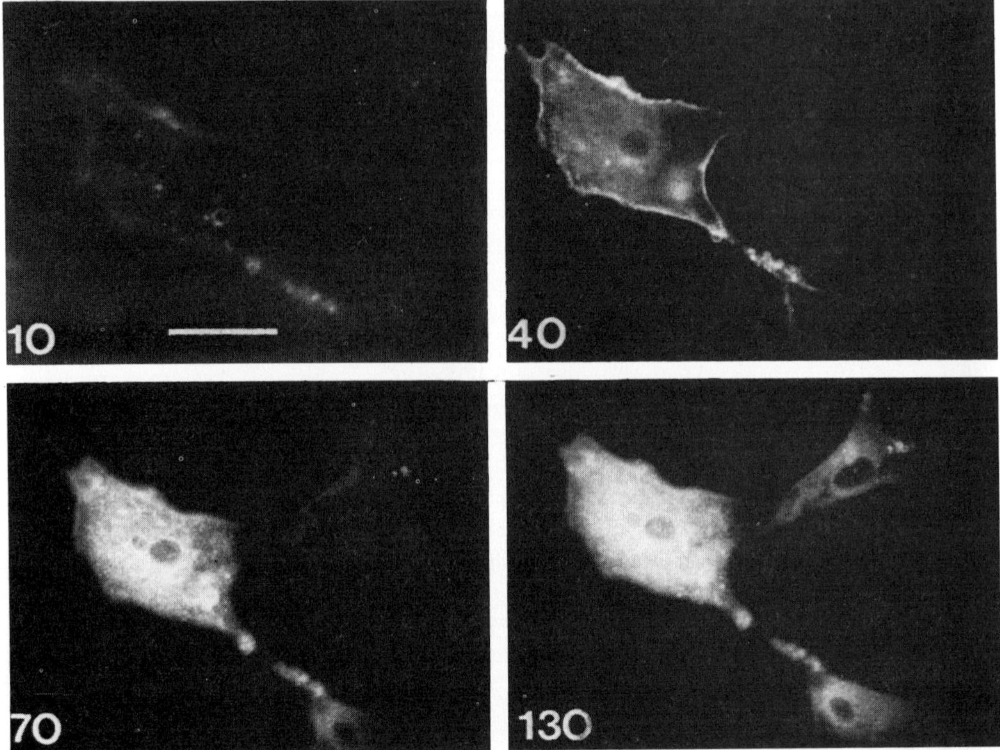

Fig. 4.6 The uptake of Evans Blue into chick fibroblasts. A series of micrographs taken at the time intervals (secs) shown after exposure to a 0.01% solution of the fluorochrome. The fluorescence is initially associated with the cell surface, then enters the cytoplasm, where it eventually becomes localized in granules. The bar is 30-μm long

Evans Blue seems initially to bind to the cell surface and then be internalized into granules which may represent lysosomes. As this hydrophilic fluorochrome is known to bind avidly to protein it appears likely that it gains entry via such binding to membrane proteins. That the increase in measured cytoplasmic fluorescence only occurs for a limited time may indicate that such an uptake process can quickly be saturated. These data are clearly in keeping with what is known about endocytosis.

This work is being extended to determine whether other substances which are known to be retrogradely transported are taken up in such a way, and whether neuronal cells, grown in tissue culture, behave similarly to fibroblasts.

The next problem relates to investigating the process of retrograde axonal transport of exogenous molecules. How can this be carried out in cell culture? Here it is necessary to set up a system in which it is possible to expose the axon terminals to the fluorescent substance under investigation but keep the cell bodies separate. Thus, if the cell bodies acquire the fluorochrome then the only route will be transport via the axon. Campenot (1977), and since then other authors (Chan *et al.*, 1981; Thoenen and Schwab, 1983), has shown that multi-compartment cell culture systems can be used in this way.

Neurones in a central cell culture compartment will, given appropriate conditions, extend neurites through a thin, separating layer of silicone grease into outer compartments provided they contain a suitable growth medium. A similar, rather simpler, system has been set up in our laboratory (Fig. 4.8). Two concentric plastic rings are sealed to a collagen-coated glass coverslip by high quality (Dow-Corning) silicone vacuum grease. Neuronal cells (and growth medium) are introduced into the central compartment and allowed to adhere to the substrate. After some hours in culture they extend neurites from their cell bodies (Fig. 4.9); if these grow through the grease seal then the fluorescent substance under investigation is introduced into the outer compartment. If retrograde transport of such a fluorochrome occurs then it will, after an appropriate time, label neuronal cell bodies in the central chamber. The time course and amount of transported material can be investigated relatively easily. It is, of course, necessary to do careful control experiments to ensure that there is no leakage of material from outer to inner compartments.

Thus, in an experimental system which does not use whole animals, the phenomenon of

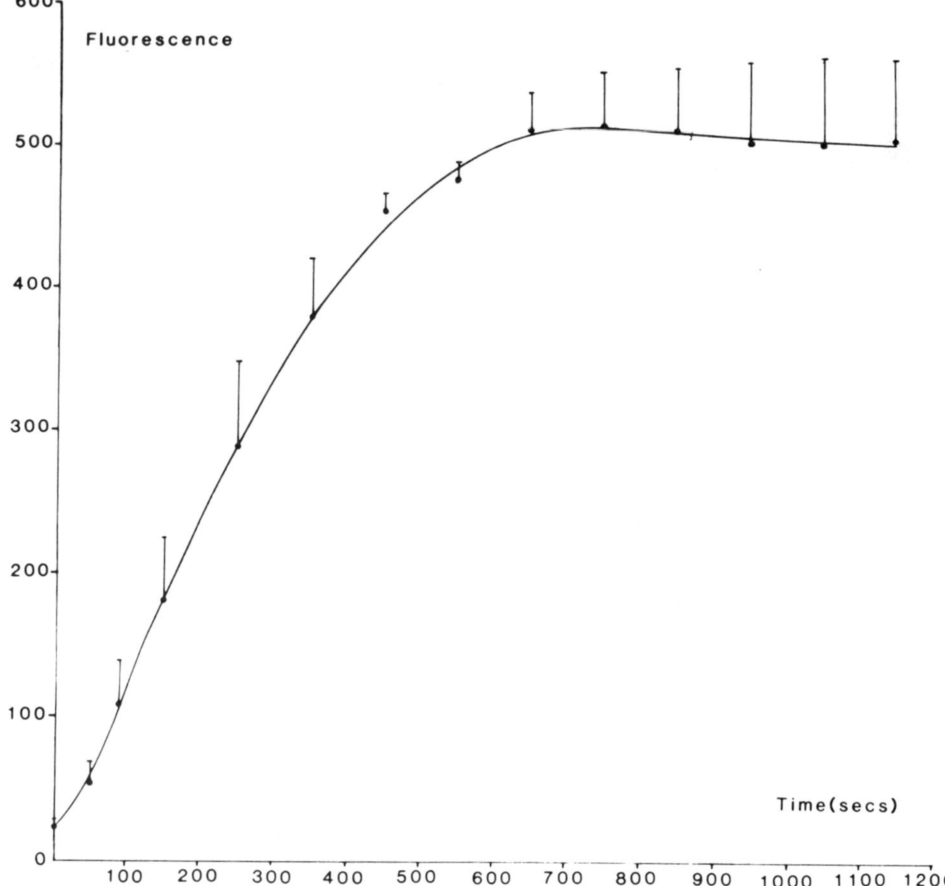

Fig. 4.7 The change in intensity of fluorescence in the cytoplasm, measured photometrically, during exposure of chick fibroblasts to 0.01% Evans Blue. The graph shown represents the averaged results from four such cells

retrograde axonal transport can be investigated. This allows the screening of a large number of compounds to determine what physico-chemical properties those which are taken up and transported have in common. As mentioned earlier, this will have important consequences not just for our understanding of axonal transport but also has particular relevance in clinical medicine.

Finally, the advantages of these cell-culture methods over those involving whole animals should be discussed. The first is, of course, the humanitarian consideration that this approach obviates the need for living animals to have to undergo experimental surgery and later to be killed. Naturally the cells used in tissue culture methods have originally to come from whole living organisms but an extremely large number of experiments can be done with the cells (and their daughter cells grown in culture) from one animal compared with the usual ratio of one experiment to one animal using a more traditional approach.

There are, however, more disadvantages with the use of whole animals. Firstly, it is clear that tissue culture allows a more precise control of the experimental system. Thus, for example, it is possible to expose the cells or parts of cells to a known concentration of fluorochrome rather than attempt to guess the effective uptake concentration when such a substance is injected into the nervous system of a whole animal.

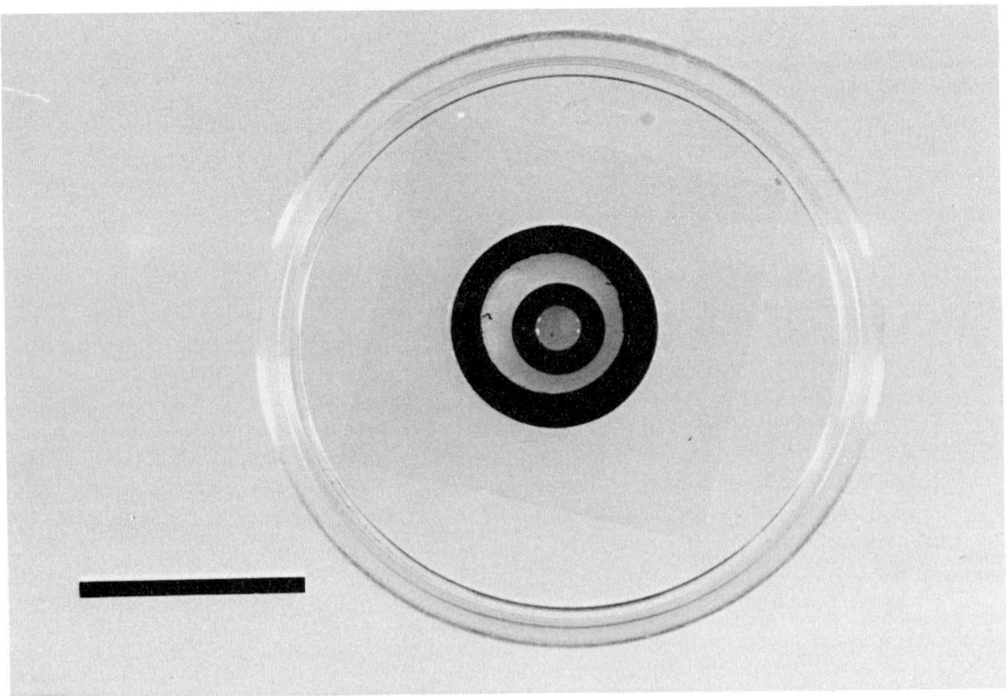

Fig. 4.8 The culture system, consisting of two concentric compartments, used for investigation of retrograde transport of exogenous compounds in neurones in cell culture. The two black plastic rings are sealed to the collagen coated coverslip by a thin layer of silicone grease. The bar is 20-mm long

Secondly, if a whole animal is the experimental unit it is obviously more expensive in terms of time and cost to do enough experiments to evaluate the effects of inter-animal variation. It is much easier to use individual cells as the experimental unit. Moreover, many important phenomena which are occurring at the cellular level may be missed when the averaged results from many cells are observed in a whole animal system.

Thirdly, it is known that general anaesthetics, especially barbiturates, have an inhibitory action on uptake and axonal transport. One has to use anaesthesia in whole animal work so it is quite possible that the result may be affected by the fact that anaesthesia was used when injecting the substance to be investigated. In cell culture, of course, this problem is completely avoided.

Finally, when cells are investigated in tissue culture the whole cell is present. It is often forgotten that if microscope sections are cut from a piece of tissue from a whole animal many of the cell profiles seen will be only parts of cells. It is usually difficult or impossible to tell whether a whole cell is present or not. This has important implications if one is trying to make measurements from such cells.

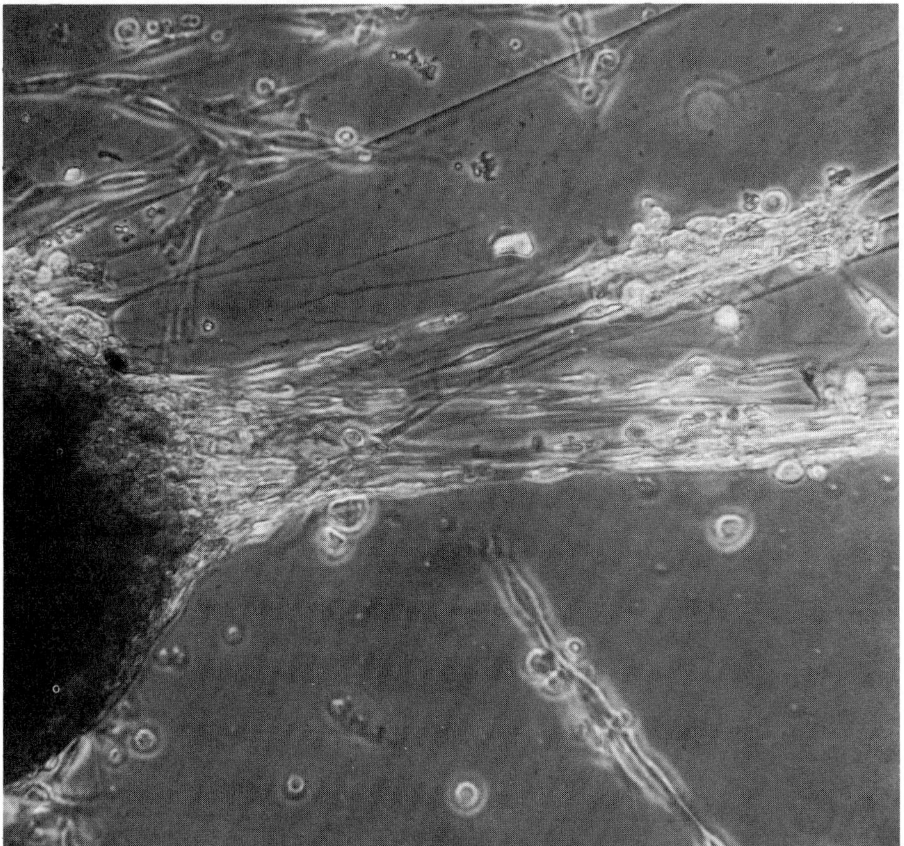

Fig. 4.9 A cluster of neurones which have grown lengthy neurites in the cell culture system used. (× 300)

Obviously, there are also disadvantages in a tissue culture system. It is never possible completely to mimic the conditions *in vivo*, but it is important to realize that an experimental system that involves the use of whole animals may not be as representative as one would like either.

Acknowledgement

The generosity of the Humane Research Trust allowed purchase of the laminar flow chamber and incubator in this study. The Trust not only allowed the cell culture work to commence but also stimulated the author's interest in alternatives to experimental techniques using living animals.

References

Campenot, R. B. (1977). Local control of neurite development by nerve growth factor. *Proc. nat. Acad. Sci. USA*, **74**, 4516–19.

Chan, K. Y., Bunt, A. H. and Haschke, R. H. (1981). *In vitro* retrograde neuritic transport of horseradish peroxidase isoenzymes by sympathetic neurons. *Neuroscience*, **6**, 59–69.

Forman, D. S. (1982). Microscopic methods for the observation of axonal transport in living axons. In *Axoplasmic Transport*. Ed. Weiss, D. G. Berlin, Springer-Verlag, pp. 424–8.

Hamperl, H. and Schümmelfeder, N. (1961). Some aspects of fluorescence microscopy. *CIBA Symposium*, **9**, 50–63.

Haschke, R. H., Ordronneau, J. M. and Bunt, A. H. (1980). Preparation and retrograde axonal transport of an antiviral drug/horseradish peroxidase conjugate. *J. Neurochem.*, **35**, 1431–5.

Hendry, I. A. and Campbell, J. (1976). Morphometric analysis of rat superior cervical ganglion after axotomy and nerve growth factor treatment. *J. Neurocytol.*, **5**, 351–60.

Innocenti, G. M. (1981) Growth and reshaping of axons in the establishment of usual collosal connections. *Science*, **212**, 824–27.

Jakobsen, J., Sidenius, P. and Brændgaard, H. (1986). A proposal for a classification of neuropathies according to their axonal transport abnormalities. *J. Neurol. Neurosurg. Psychiat.*, **49**, 986–90.

Koda, L. Y. and Van der Kooy, D. (1983). Doxorubicin: a fluorescent neurotoxin retrogradely transported in the central nervous system. *Neurosci. Lett.*, **36**, 1–8.

Kristensson, K. (1970a). Transport of fluorescent protein tracer in peripheral nerves. *Acta neuropath.*, **16**, 293–300.

Kristensson, K. (1970b). Morphological studies of the neural spread of herpes simplex virus to the central nervous system. *Acta neuropath.*, **16**, 54–63.

Kristensson, K. and Olsson, Y. (1971). Retrograde transport of protein. *Brain Res.*, **29**, 363–5.

Kuypers, H. G. J. M., Catsman-Berrevoets, C. E. and Padt, R. E. (1977). Retrograde axonal transport of fluorescent substances in the rat's forebrain. *Neurosc. Lett.*, **6**, 127–35.

Ochs, S. (1972). Fast transport of material in mammalian nerve fibres. *Science*, **176**, 252–60.

Ploem, J. S. (1970). Standards for fluorescence microscopy. In *Standards in Immunofluorescence*. Ed. Holborow, E. J. Oxford, Blackwell Scientific Publications. pp. 137–54.

Thoenen, H. and Schwab, M. (1978). Physiological and patho-physiological implications of retrograde axonal transport of macromolecules. *Advances in Pharmacology and Therapeutics*, Volume 5: *Neuropsychopharmacology*. pp. 37–59. Oxford and New York, Pergamon Press.

Thoenen, H. and Schwab, M. (1983). Mechanism of uptake and retrograde axonal transport of noradrenaline in sympathetic neurons in culture; reserpine-resistant large dense-core vesicles as transport vehicles. *J. Cell Biol.*, **96**, 1538–47.

CHAPTER 5

The acquisition and differentiation of oligodendrocytes in re-aggregating glial cell cultures

E. M. Carey, M. Vojvodic, R. Reynolds and N. Herschkowitz

Glossary of Terms

Abbreviations used in text:

AC	astrocytes
cAMP	3′, 5′-cyclic adenosine monophosphate
CNP	2′, 3′-cyclic nucleotide 3′-phosphohydrolase
CNS	central nervous system
Cyt Ara	cytosine arabinoside
DMEM	Dulbecco's modified Eagles' medium
EGF	epidermal growth factor
div.	days *in vitro*
ELISA	enzyme-linked immunosorbent assay
E.M.	electron microscopy
FITC	fluorescein isothiocyanate
GC	glial cells
GFAP	glial fibrillary acidic protein
GS	galactocerebroside
IGF	insulin-like growth factor
MBP	myelin basic protein
ODC	oligodendrocytes
PAGE	polyacrylamide gel electrophoresis
PBS	phosphate buffered saline
PMA	phorbol myristate acetate
S^+	serum containing
S^-	serum free
SDS	sodium dodecyl sulphate
T–TBS	Tween 20–Tris buffered saline

Introduction

The control of cell proliferation and differentiation in the CNS is critical for normal brain development. Endocrine and paracrine agents orchestrate the acquisition of the different types of neural cells in a precise time schedule and promote cell differentiation by the expression of phenotypic cell-specific components and morphology.

Our interest in the process of myelination in the CNS originated from our observations that hypomyelination in the normally heavily myelinated corpus callosum region of the infant CNS frequently occurred in cases where death was associated with respiratory insufficiency due to premature birth and congenital heart defects, but also in nearly half the cases of Sudden Infant Death (Carey and Foster, 1984; Foster and Carey, 1983). The main function of peri-neuronal oligodendrocytes (ODC), which are one type of glial cell present in myelinated white matter of the CNS, is the production of multi-lamellar myelin which surrounds the nerve axon in order to promote rapid conduction of the nerve action potential. We found that the deficit in myelin was primarily due to a reduction in glial cell numbers, because the average amount of myelin produced per cell appeared to be the same as in the brain of infants dying from unrelated illnesses. We suggested that the insult to glial cell acquisition occurred around the time of birth and before the later period of myelination. In animal studies, undernutrition and hypobaric oxygen exposure also give rise to hypomyelination. A deficiency of growth hormone in Snell dwarf mice, which congenitally lack growth hormone, or in rats made deficient by injection of antibody to growth hormone at birth, causes hypomyelination and a corresponding reduction in glial cell numbers. Normal myelination in the developing brain therefore requires the proliferation of glial precursor cells and their differentiation into myelin-producing ODC cells at a critical period of development. The integrated responses to endocrine and paracrine agents regulate both cell proliferation and differentiation.

We have studied the formation of ODC in primary cultures of cells dissociated from neonatal rodent brain, which form surface-adhering mixed glial cell cultures (McCarthy and de Vellis, 1980; Bologa *et al.*, 1981; Ecclestone and Silberberg, 1984). Most glial precursor cells differentiate into epitheloid-type astrocytes (AC) which proliferate rapidly to form in time a confluent monolayer. Process-bearing oligodendrocytes and undifferentiated cells grow on the AC layer, but without forming tight junctions which are visible by electron microscopy between AC cells. Few neuronal cells survive the mechanical dissociation, probably because in the neonatal brain neurones are already differentiated. Surface-adhering cell cultures are becoming widely used as a model to study developmentally related processes in neurobiology because of the ease with which the concentration of hormone or growth factor can be precisely regulated in a chemically-defined culture medium, with respect to concentration and time of exposure. Phenotypic antigens expressed at various stages of differentiation can be detected using immunocytochemical methods with good penetration of specific antibodies reacting with cell surface or intracellular antigens, without loss of characteristic cell structure. A combination of morphological and immunocytochemical criteria can be used to distinguish different types of brain cells and quantitate the density of specific cell types. However, surface-adhering cell growth is essentially two dimensional, and cell morphology and cell–cell contacts may develop differently from the situation *in vivo*. An alternative to primary glial cell cultures would be the use of established cell lines, but although a number of oligodendrocytoma cell lines have been characterized, transformed cells do not show all the characteristic features of the terminal differentiation of ODC cells.

Here, we describe some aspects of the differentiation of ODC in primary glial cell cultures. We have examined: (i) the proliferation of cells in a chemically-defined medium, compared with serum-containing medium, to which growth factors have been added; (ii) cell morphology and the formation of myelin-like membranes; (iii) the co-ordinated expression of myelin proteins during cell differentiation; and (iv) the synergism between insulin and epidermal growth factor (EGF) in promoting ODC formation.

Experimental procedures

Mixed glial cultures were prepared from neonatal mouse forebrain according to the method of Bologa *et al.* (1981). Cells were mechanically dissociated by trituration into Dulbecco's modified Eagle's medium (DMEM) containing 10% foetal calf serum, supplemented by the addition of 30 mM D-glucose, 200 nM L-glutamine and 100 units/ml penicillin and 75 μg/ml streptomycin (S^+ medium). Viable cells were plated at a density of 2.5×10^6 cells /cm^2 in 2.0 ml medium into 35-mm diameter wells of plastic culture plates (Falcon multi-well) pre-coated with poly-L-lysine. Cells were cultured at 37°C in 5% CO_2 and 80% humidity. Medium was replaced completely after 3 days with either fresh serum-containing (S^+) medium or a chemically-defined, serum-free (S^-) medium (Ecclestone and Silberberg, 1984). This consisted of equal volumes of DMEM and Ham's F12 media supplemented with glucose and glutamine (as in the S^+ medium), together with insulin (10 μg/ml), transferrin (50 μg/ml), putrescine (100 μM), selenium (30 nM) and tri-iodothyronine (15 nM). Further additions are indicated below. Subsequently, the culture medium was replaced every 3 days. Cultures were carried out in triplicate.

Before staining, cells were first washed in Hank's BSS. For the detection of cell-surface galactocerebroside (GC), cells were exposed to appropriately diluted rabbit antiserum to galactocerebroside for 30 min, washed and then exposed to fluorescein isothiocyanate (FITC)-conjugated goat anti-rabbit IgG (Cappel), washed and fixed in 4% formaldehyde in phosphate-buffered saline (PBS), covered with 90% glycerol in PBS and mounted with a glass cover slip. For the detection of intracellular antigens, cells were washed and fixed in formaldehyde followed by 90% acetone (−20°C) for 1 min. Cells were then incubated with 5% normal goat serum followed by exposure to primary and secondary antibodies diluted in PBS buffer. Between each step cells were washed with PBS. Cells were viewed with a Leitz Fluovert inverted microscope.

Biochemical assays (quantitation of DNA, protein, 2′3′-cyclic nucleotide-3′-phosphohydrolase (CNP) enzyme activity and myelin proteins) were carried out after cell disruption by sonication (60 sec at 50 W using the microprobe of a Branson sonicator) in 2.0 ml 0.01% sodium dodecyl sulphate (SDS) in PBS. Duplicate portions from each culture were taken for the estimation of protein and DNA by standard procedures. The remainder was used to quantitate cell-specific antigens by a solid phase enzyme-linked immunosorbent assay (ELISA). Solubilized cell protein (or purified myelin basic protein (MBP) or CNP antigen (1 μg/ml) or myelin membrane proteins (2.5 μg/ml) in PBS containing 0.01% SDS and 250 μg/ml serum albumin) was bound to 8 mm nitrocellulose discs by immersion of discs in the protein solution (5 discs/ml) until saturated with protein (from decrease in absorbance at 280 nm). After drying, any unreacted sites were blocked in 50 mM Tris/HCl buffer, 0.154 M NaCl, pH 7.6, containing 0.5% Tween-20 (T–TBS buffer) for 1 h, followed by washing and storage of discs immersed in water at 4°C until used for assay. Antisera, or normal rabbit serum as control, diluted in 0.05% T–TBS (1:

2000 fold or greater) were incubated with individual discs in each well (0.25 ml/well) or a 24-well Linbro plate for 3 h. Quantitation of bound antibody was carried out using the procedure of Blake *et al.* (1984), except that alkaline phosphatase activity of goat anti-rabbit IgG second antibody conjugate (Tago Inc.) was determined using *p*-nitrophenylphosphate (1 mg/ml) in 0.1 M veronal buffer as substrate. Standard curves were constructed using purified antigens and the amount of unknown antigen was determined directly. Antibody titres of the anti-CNP and anti-MBP antibodies (prepared as described below) were also determined using the solid phase ELISA. The enzyme activity of CNP was assayed using 2′, 3′-cyclic NADPH as substrate. One unit of activity is equivalent to one micromole substrate hydrolysed per minute.

Oligodendrocytes were identified by immunostaining with polyclonal antibodies to galactocerebroside, CNP and MBP. A rabbit polyclonal antibody to mouse myelin CNP was prepared by incorporating the purified protein into single bilayer phosphatidylcholine liposomes (Mimms *et al.*, 1981). Rabbit anti-bovine CNP antibody (Kim *et al.*, 1984) was provided by Dr Arthur McMorris of The Wistar Institute, Philadelphia. Rabbit anti-MBP antibody was produced to the 18-kD component of rabbit MBP and purified by affinity chromatography. Astrocytes were identified by immunostaining with a polyclonal rabbit antibody to glial fibrillary acidic protein, GFAP (Dako Corp.). Rabbit antibody to rat brain γ-enolase was purchased from Polysciences Inc. Characterization of antibodies to MBP and CNP proteins was carried out by immunodetection of myelin proteins on Western blots after SDS–PAGE (Blake *et al.*, 1984), and proteins were visualized in the gel by silver (Merril *et al.*, 1981), or Coomassie staining.

Results

2′, 3′-Cyclic nucleotide-3′-phosphohydrolase (CNP) activity is present in myelin and myelin-related membranes, and in plasma membranes of isolated ODC, although its functional significance is unclear. MBP proteins are major components of myelin. We therefore used antibodies to these proteins to identify ODC in glial cultures. Enzyme activity (for CNP) and immunoreactive protein content of cultures were quantitated as parameters reflecting ODC differentiation and myelin-like membrane formation. Two polyclonal anti-CNP antibodies to mouse and bovine CNP were used. Both antibodies react in an identical manner with a protein doublet in the Wolfgram protein region of myelin (the W1a and W1b, according to the terminology of Kim *et al.*, 1984) on electroblots of mouse myelin. The anti-mouse CNP antibody was capable of aggregating membrane vesicles obtained by sonication of mouse or bovine myelin, while the anti-bovine CNP could not, suggesting that anti-mouse CNP antibody reacts with epitopes of CNP protein which are exposed in myelin. The anti-MBP antibody reacted with large (18-kD) and small (14-kD) MBP proteins, and pre-large and pre-small proteins.

In the mixed glial cell cultures, epitheloid AC cells were identified by staining with antibody to the intermediate filament glial fibrillary acidic protein (GFAP), which reacts with cytoskeletal elements within AC cells. Most $GFAP^+$ cells are flattened, epitheloid cells (Fig. 5.1a), described by Raff *et al.* (1983a) as type-1 cells (analogous to protoplasmic astrocytes *in vivo*). In mixed glial cultures from the frontal cortex a small number (usually 1–2%) of $GFAP^+$ AC cells have a stellate morphology. These type-2 AC cells are more abundant in optic nerve cultures (Raff *et al.*, 1983a), and may be analogous to fibrous astrocytes which are localized around nerve fibres *in vivo*. By 10 days *in vitro* (div.), AC cells form a near confluent cell layer. Phase dark cells grow on the AC layer, and CNP$^+$ ODC cells can be detected on the AC cell surface from around 7 div. in serum-free medium (Fig. 5.1b). Most CNP$^+$ cells elaborate a fine network of processes, and at higher magnification

a

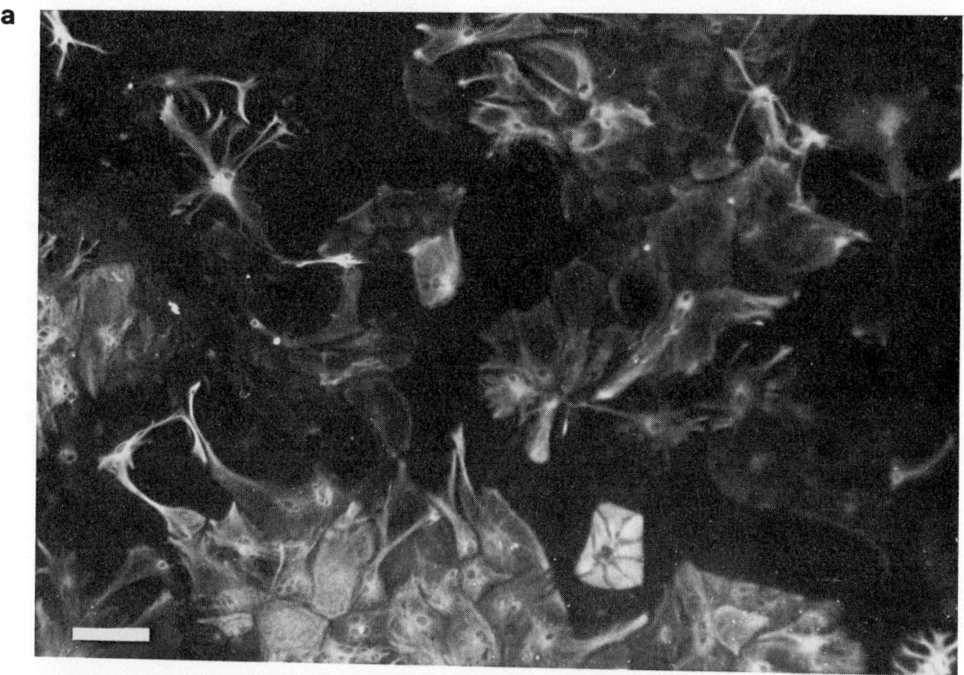

b

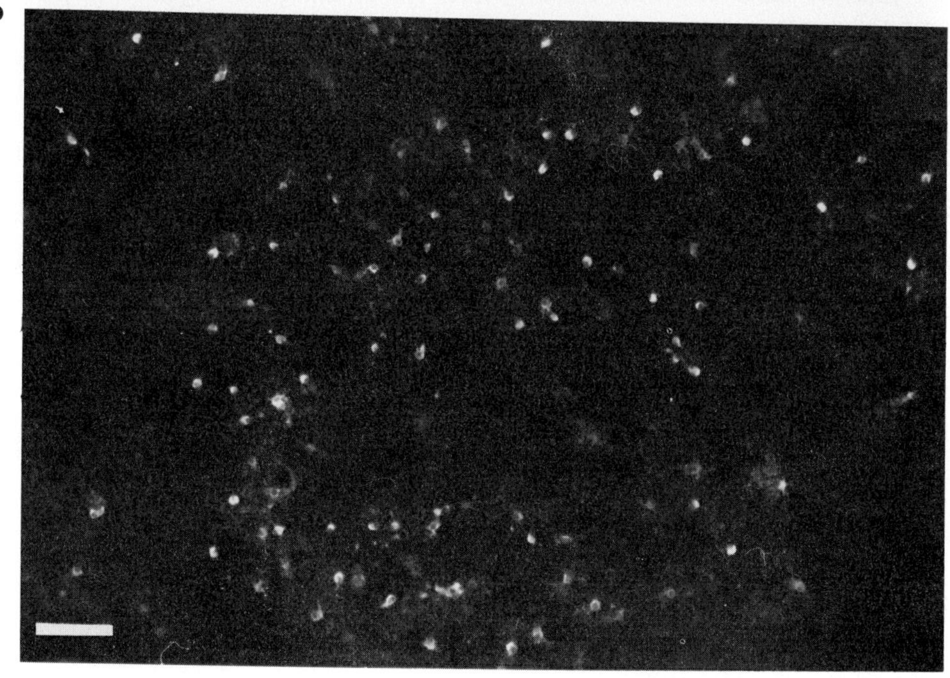

c

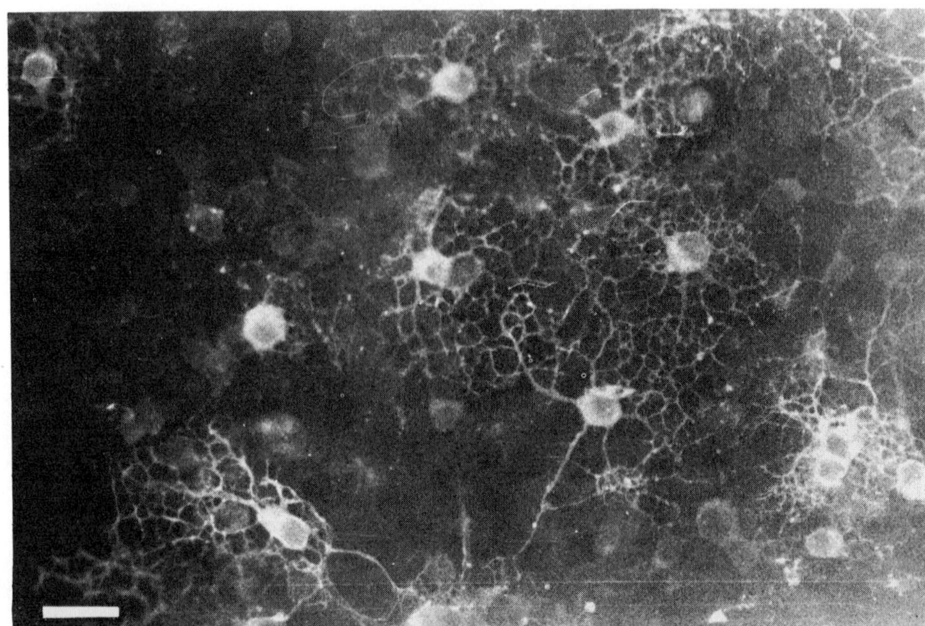

Fig. 5.1 Indirect immunofluorescence staining of cells in 10 days *in vitro* (div.) mixed glial cultures grown in serum-containing medium for 3 days followed by transfer into serum-free DMEM/Ham's medium: (a), culture stained with anti-GFAP antibody (note the presence of mainly epitheloid type-1 AC, and small numbers of stellate type-2 AC); (b) and (c), identical culture stained with anti-mouse CNP antibody at low and high magnification (note the fine interconnecting and overlapping network of CNP^+ processes of ODC). Scale bar represents 150 μm in (a) and (b), 40 μm in (c)

Fig. 5.2 Top: A higher magnification micrograph of cells from a culture maintained for 10 div. in serum-free medium, and stained with anti-mouse CNP antibody. Bottom: corresponding phase micrograph. Note the large CNP^+ cell forming a circular structure on the surface of AC cells and enclosing an unstained phase-dark cell. The heterogeneity of the size of processes and cell morphology is apparent. Some phase-dark cells are weakly stained and have few processes. Scale bar represents 40 μm

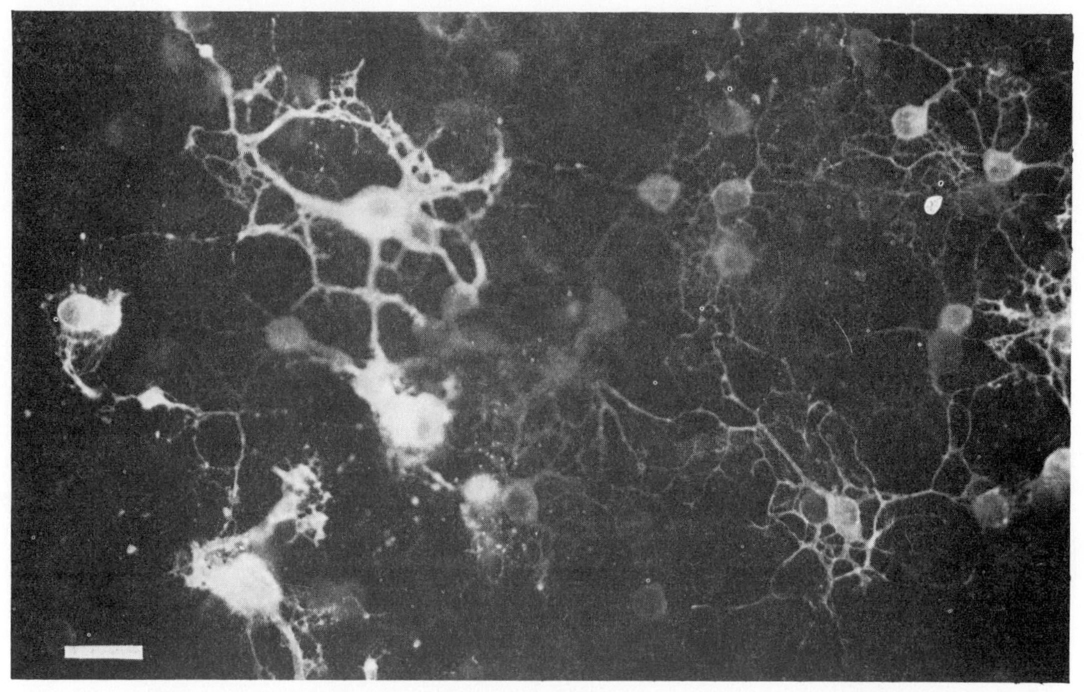

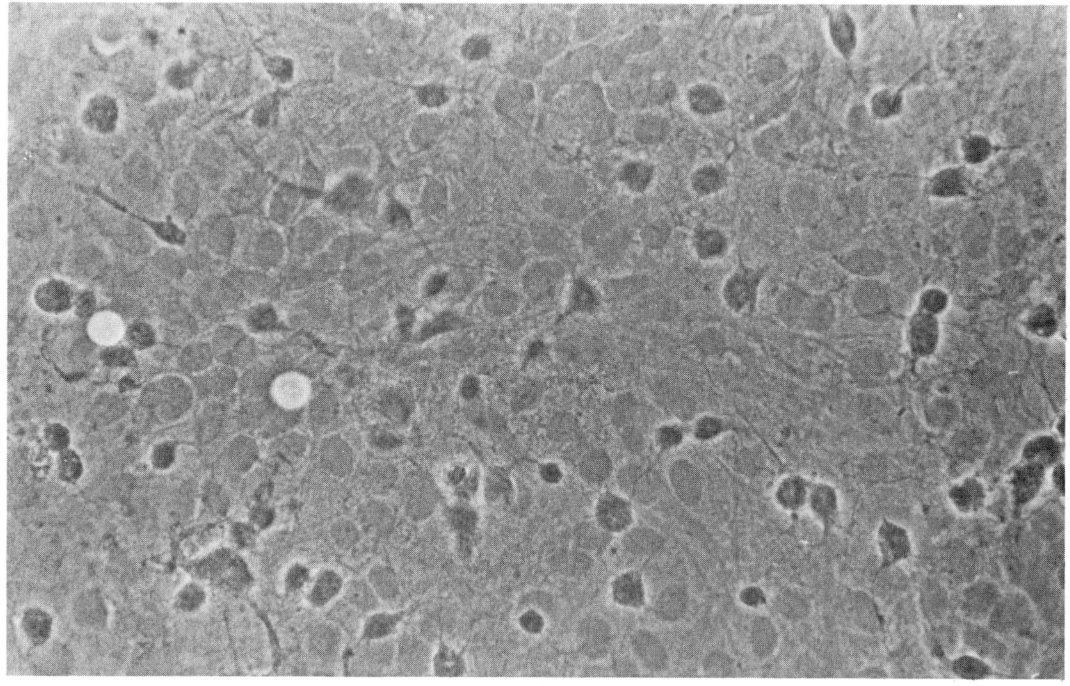

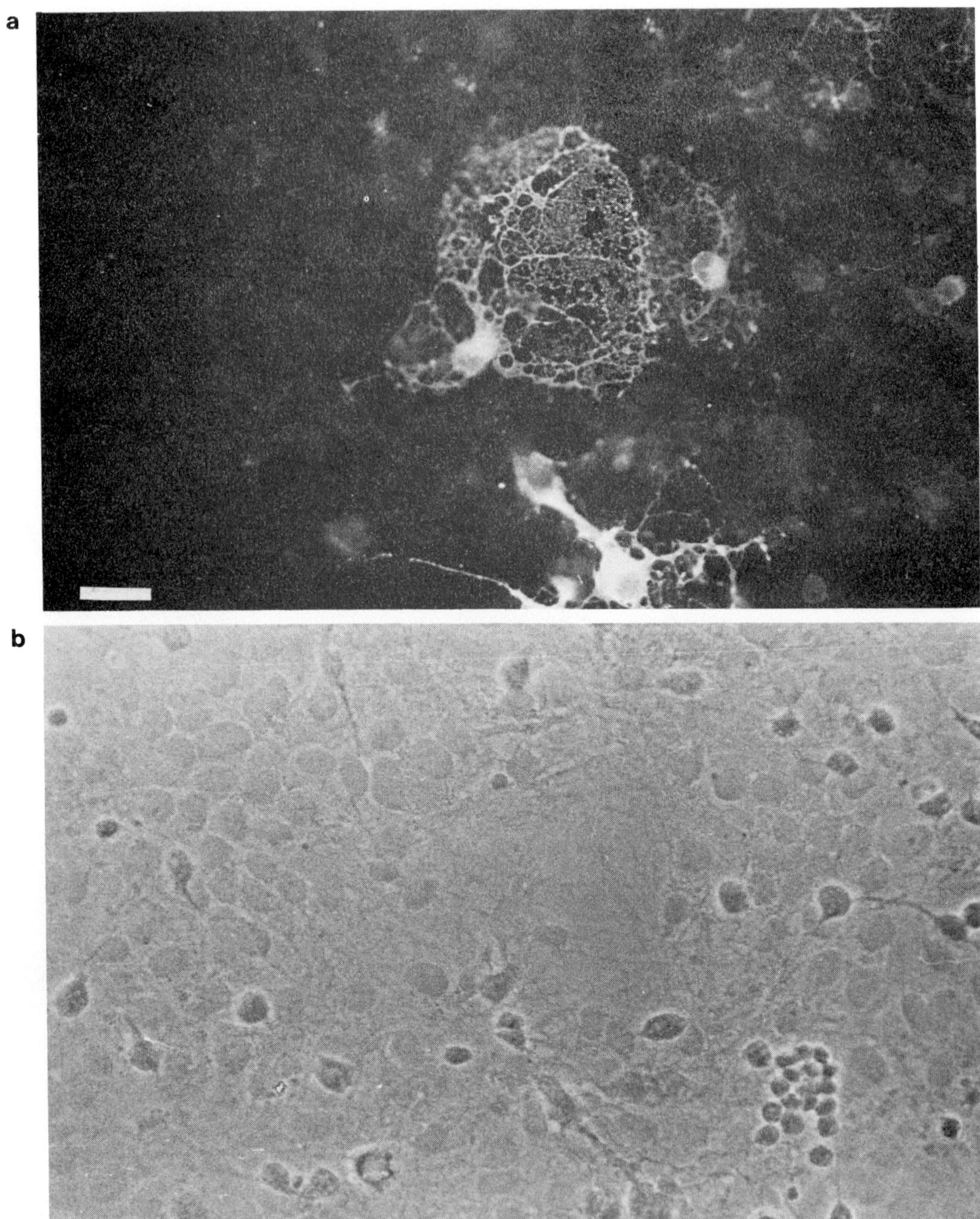

Fig. 5.3 Indirect immunofluorescence staining with anti-mouse CNP antibody of ODC in mixed glial cell cultures maintained in serum-containing medium for 14 div. A mature, terminally-differentiated ODC is shown in (a) and other stained cells show different degrees of process formation; corresponding phase contrast photomicrograph shown in (b). Circularization

c

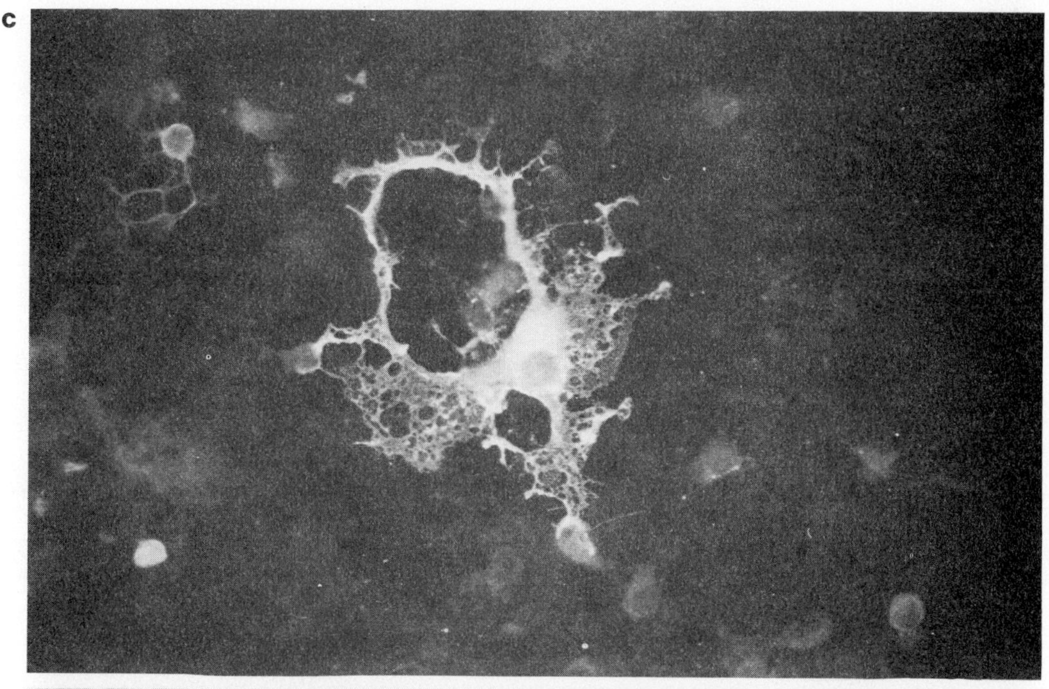

d

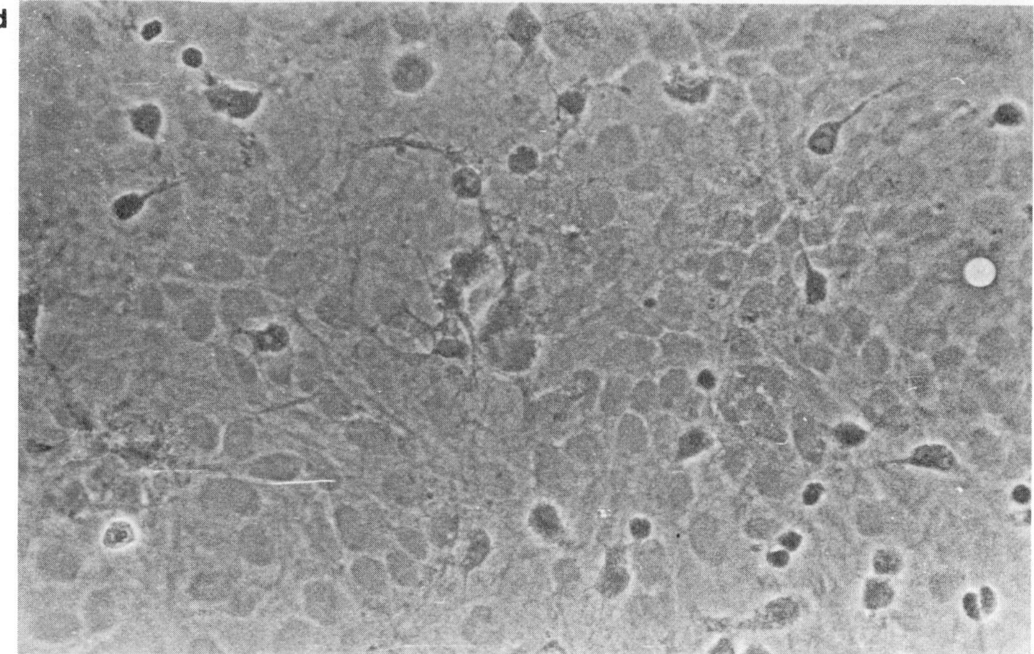

of major processes are a frequently observed structural feature as shown in (c) and processes enclose unstained phase-dark cells, seen in the phase micrograph (d). About 25% of phase-dark cells on the underlying AC cell layer are CNP^+. Fewer CNP^+ cells are evident in S^+ medium compared with S^- medium (see Fig. 5.1). Bar represents 40 μm

(Figs. 5.1c and 5.2) staining for CNP can be seen in the cell body, and in a fine network of interconnecting processes. In serum-containing medium compared with serum-free medium, a greater proportion of CNP^+ cells elaborate sheets of membrane with interconnecting processes. Thicker processes emanating from the cell body often can be seen to form circular structures (Fig. 5.3). There are always a number of phase-dark cells on the AC layer which do not stain with antibody to ODC, and some of these cells may be undifferentiated glial precursor cells. By 14 div., 40–50% are GC^+, 20–25% are CNP^+ and 5–10% are MBP^+ cells. Whether the phase-dark cells are bipotential glial progenitor cells (Raff *et al.*, 1983a; Mercanti *et al.*, 1987) remains to be established.

We examined the morphology of ODC in more detail by isolating loosely-attached precursor cells from mixed glial cultures by the shake-off procedure of McCarthy and de Vellis (1980). After shake-off, glial progenitor cells re-attach and differentiate mainly into ODC (Reynolds *et al.*, 1988). We found that in these ODC enriched cultures, CNP was localized in the cell body, major processes, the fine interconnecting network of processes in the flattened membrane extensions and at the extremities of the flattened membranes (Fig. 5.4a). MBP had an opposite location, being present in the membrane of the flattened sheets but not in the major or interconnecting fine processes (Fig. 5.4b). It has previously been found that CNP and MBP proteins are present in bulk-isolated ODC from the CNS (Szuchet and Stefansson, 1980), and in membrane processes of differentiating ODC in culture (Roussel *et al.*, 1983; Sprinkle *et al.*, 1983; Kim *et al.*, 1984). At different stages of differentiation, proteins first appear in the cell body and later in cell processes. In culture, the production of myelin-like membrane expansions may be a characteristic of mature, terminally-differentiated ODC. The relationship between the appearance of MBP and the formation of myelin-like processes by ODC will be discussed later. Whether there are structural filaments containing the W1 proteins (recognized by the anti-CNP antibody) in myelin produced *in vivo* is unknown because of the experimental difficulty of obtaining penetration of antibody into multi-lamellar myelin. In culture, ODC cells did not form lamellar membrane structures, and it may be that in the absence of neurones, membrane sheets fail to wrap around other cell structures. However, lamellar structures observed by E.M. have been described.

Since the extending network of processes was a function of ODC differentiation, we were interested to see if cell morphology was altered by exposure to phorbol myristate acetate (PMA), which is a known tumour promoter. Exposure of cultures at 3 div. to 25 nM PMA for 1–5 div. resulted in the formation of long membrane processes (Fig. 5.5). Immunofluorescence staining (Figs. 5.5a and 5.5c) revealed the presence of CNP and MBP in the cell body and convoluted fine processes of ODC. Thinner processes showed a blebbing or thickening at regular intervals. The most noticeable change in the disorganization of the cytoarchitecture was the absence of myelin-like membrane sheets. The reason for this is unknown, but one possibility is that PMA causes an alteration in the structural organization of cytoskeletal proteins as a result of increased protein phosphorylation brought about by the activation of a calcium/phospholipid-dependent protein kinase C.

Foetal calf serum contains a number of mitotic growth factors, such as platelet and placental-derived growth factors, which stimulate the proliferation of many different types of cells. The disadvantage of serum-containing media is the heterogeneity and unknown concentrations of growth factors. We, therefore, investigated the differentiation of ODC and expression of CNP and MBP proteins in a chemically-defined, serum-free (S^-) medium similar in composition to that described by Ecclestone and Silberberg (1984), and compared growth of cells in serum-containing (S^+) medium. Insulin has been found to enhance ODC formation, and therefore was included in S^- medium, although at a supra-

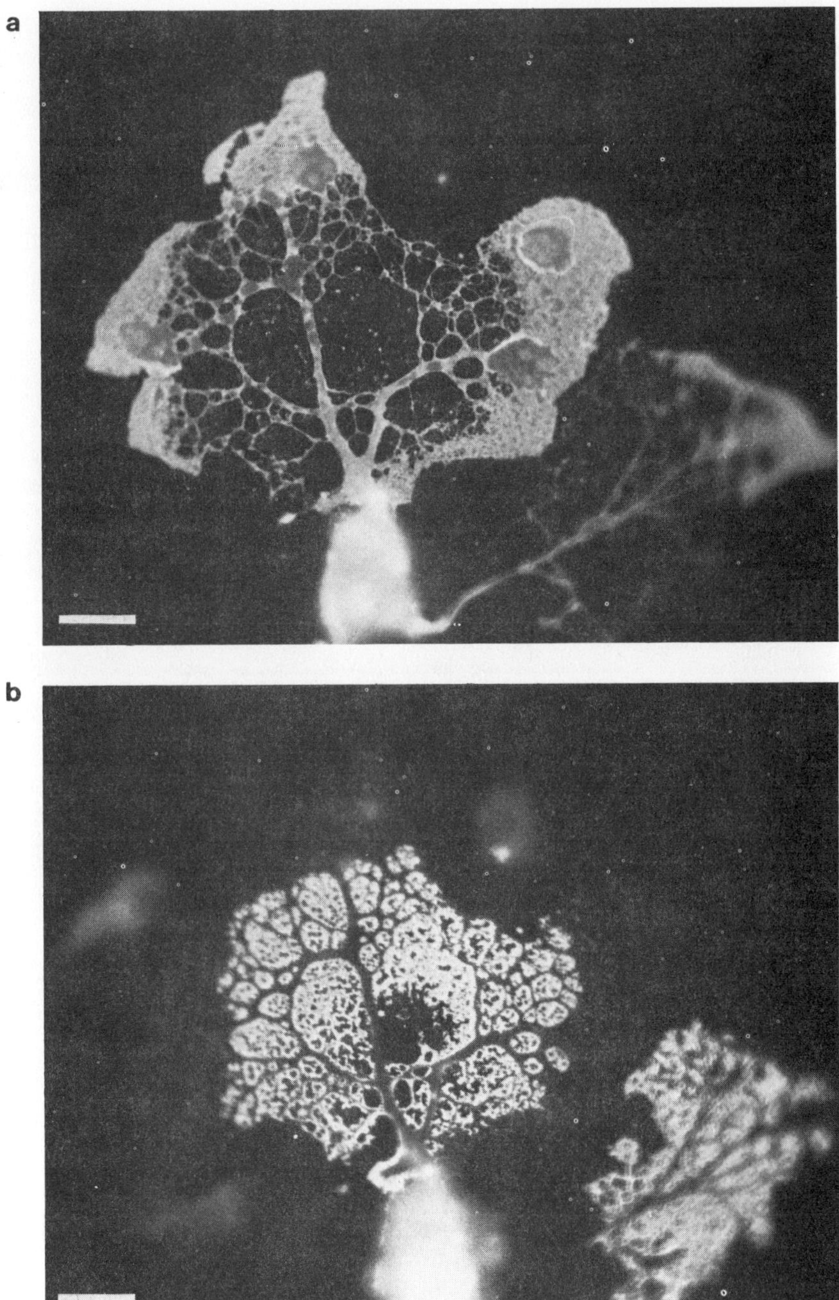

Fig. 5.4 Double immunofluorescence staining of isolated mouse ODC with (a) polyclonal anti-mouse CNP antibody and (b) mouse monoclonal anti-MBP antibody. The flat myelin-like membrane is MBP positive and CNP negative, whereas the major and interconnecting processes, and extremities of myelin-like membranes are MBP negative and CNP positive. Scale bar represents 24 μm

a

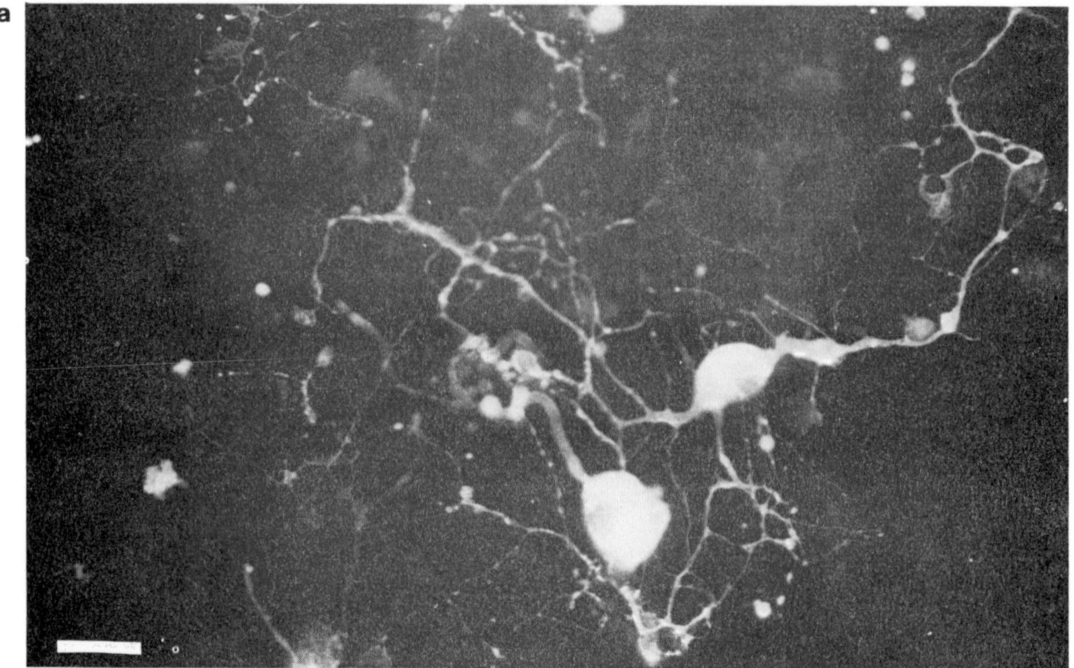

b

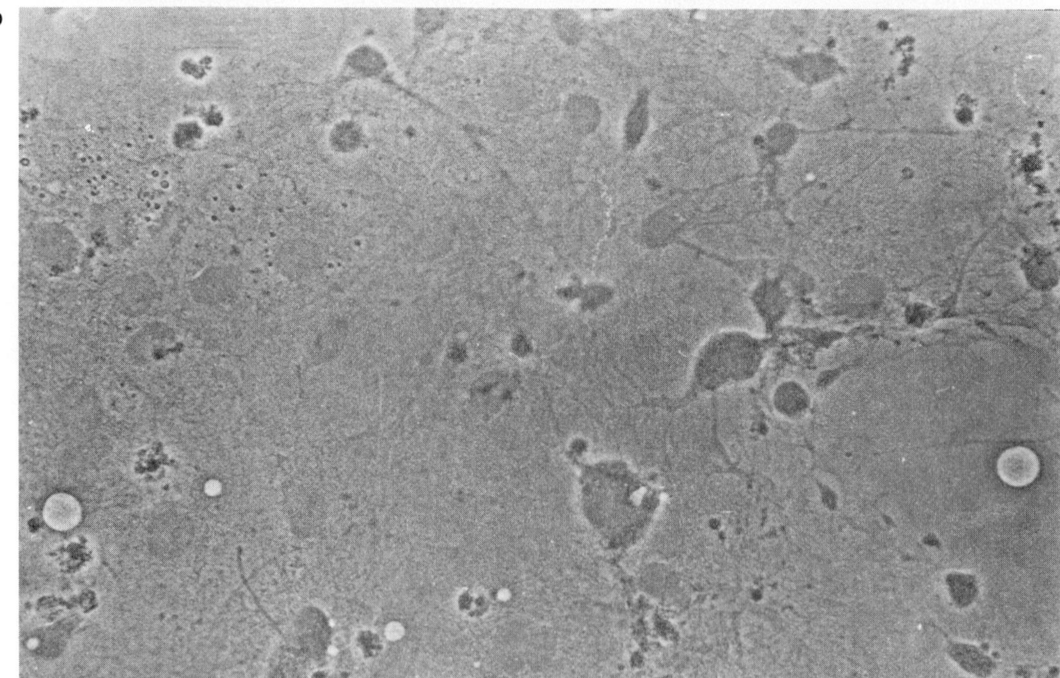

Fig. 5.5 The effect of phorbol myristate acetate (PMA) on the morphology of ODC. Cultures in serum-free medium were exposed to 25 nM PMA for 5 days at 3–8 div. and fixed at 14 div.: (a), staining with anti-mouse CNP antibody, and (b) corresponding phase micrograph; (c),

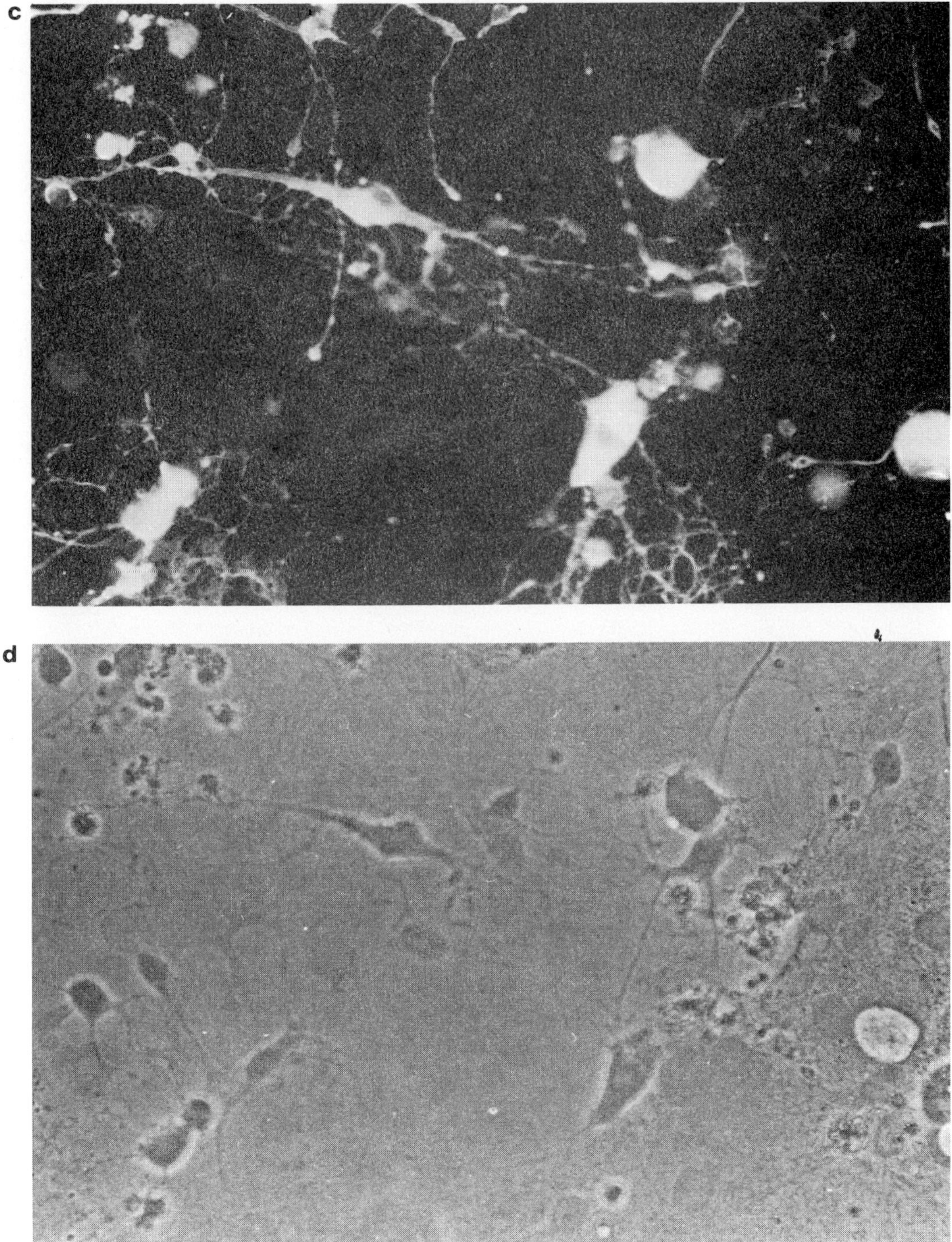

staining with anti-MBP antibody, and (d) corresponding micrograph. Note the cellular hypertrophy of ODC cell bodies, the blebbing of processes, and the absence of the fine network of interconnecting processes and myelin-like membrane sheets. Scale bar, 40 μm physiological concentration (10 μg/ml). Cells were initially maintained in S^+ medium for 3 div., during which time undifferentiated surface-adhering cells rapidly proliferated, before transfer into S^- medium from 3 div. onwards. In our recent experiments (described later) serum has been replaced by a serum substitute (Gibco UltraSer) which supplements the DMEM/HAMS medium with mitotic growth factors.

The DNA and protein content of the mixed glial cultures is shown in Fig. 5.6a. Cultures maintained in S^+ medium had a 2-fold higher content of DNA and protein than S^- cultures. A decline in both DNA and protein occurred in S^- medium from 18 div., suggesting loss of cells in older cultures. CNP^+ and MBP^+ ODC were first detected in S^- cultures from 7 div., and some cells were already elaborating membrane processes from 7 div. onwards. The number of CNP^+ and MBP^+ ODC cells in S^- medium was higher than in S^+ medium, the number of cells increased in parallel, and cell differentiation was accelerated in S^- medium. In both S^+ and S^- medium, the number of CNP^+ cells exceeded that of MBP^+ cells. However, the proportion of MBP^+ cells increased in S^+ medium, coinciding with an increase in the proportion of cells producing flattened membrane sheets characteristic of terminal differentiation. In double labelling experiments using rabbit polyclonal antibody to CNP and a monoclonal antibody to MBP, we found that all CNP^+ cells were also MBP^+. The results imply that only a proportion of CNP^+ cells are capable of expressing MBP and therefore reaching what might be described as a terminal differentiation stage. The reason for the inability of all CNP^+ cells to express MBP and produce flattened membrane sheets is not known, but is an important aspect of the terminal stage of cell differentiation.

In optic nerve cultures, Raff *et al.* (1983b) reported increased numbers of GC^+ ODC in serum-free cultures. In our mixed glial cultures, the number of GC^+ cells was always higher than that of CNP^+ and MBP^+ cells (Table 5.1), suggesting that only a proportion of GC^+ cells express myelin proteins. Most of the GC^+ cells remain as small, round cells, devoid of processes, whereas at the opposite extreme virtually all MBP^+ cells are process bearing. The numbers of ODC cells formed in S^+ medium is extremely variable, and depends on the batch of foetal calf serum used. The higher the mitogenic activity of cultures as evidenced by the proliferation of AC cells, then the fewer ODC cells could be detected, which would suggest that growth factor mitogens in the serum prevent the differentiation of ODC but not of epitheloid AC.

As described earlier, there were also differences in the morphology of ODC cells in S^- and S^+ medium. In S^- medium (Figs. 5.1 and 5.2) the major processes and the fine network of the interconnecting processes in ODC were less extensive than in S^+ medium (Fig. 5.3), and in S^- medium some cells were not process bearing, with staining confined to the cell body. In S^+ medium, more ODC elaborated the sheets of myelin-like membrane than in S^- medium.

The CNP and MBP protein content in S^- and S^+ cultures is shown in Fig. 5.6b,c. CNP enzyme activity and immunoreactive protein increased in parallel. The increase in the content of both CNP and MBP in both types of cultures was detectable from 7 div., but S^+ cultures contained greater amounts of CNP and MBP proteins. This ratio of CNP or MBP protein to CNP^+ or MBP^+ ODC cell numbers is higher in S^+ than in S^- medium. Taken as a parameter of cell size, this would fit in with morphological observations on the extensive membrane processes produced in S^+ medium. We also found that the ratio of CNP to MBP immunoreactive protein expressed by ODC in culture remained roughly constant, and this ratio was 5-fold higher than in myelin isolated from mouse brain. From their simultaneous induction, the expression of these two myelinogenic parameters appears

Table 5.1 The effect of cytosine arabinoside on the acquisition of oligodendrocytes in mixed glial cell cultures*

Age of culture (div.)	*Additions (div.)*			*Number of cells per mm²*			
				GC⁺	*CNP⁺*	*MBP⁺*	*GFAP⁺*
10	EGF (3–6)			443±85	46±10	21±6	210±65
	−EGF			132±67	10±2	2±1	139±47
	+EGF (3–6)	+CytAra (3–6)		25±7	<0.1	<0.1	8±4
	−EGF	+CytAra (3–6)		46±14	<0.1	<0.1	9±5
	+EGF (3–6)	+CytAra (6–10)		546±60	50±10	9±3	243±30
	−EGF	+CytAra (6–10)		165±60	13±4	3±1	205±17
17	+EGF (3–6)			320±60	25±4	21±5	298±44
	−EGF			120±34	20±4	14±4	186±27
	+EGF	+CytAra (6–10)		508±86	3±1	3±1	214±33
	−EGF	+CytAra (6–10)		72±26	2±1	1±0	169±21
	+EGF (3–6)		+EGF (10–13)	270±71	27±1	19±4	308±70
	−EGF		+EGF (10–13)	141±40	19±3	11±4	232±77
	+EGF (3–6)	+CytAra (6–10)	+EGF (10–13)	496±66	4±1	3±1	242±33
	−EGF	+CytAra (6–10)	+EGF (10–13)	79±22	2±0	1±0	351±79

*Mixed glial cultures were maintained in DMEM/Ham's F12 medium plus supplements including insulin (as described in Experimental Procedures) with the addition of 2% UltraSer for 0–1 div. Epidermal growth factor (EGF; 10 ng/ml and cytosine arabinoside (CytAra; 1 μg/ml) were added for the periods indicated in brackets. Values are the mean from at least 3 separate cultures ± SEM.

to be coordinately regulated during the differentiation of ODC in culture. The lower ratio of MBP relative to CNP compared with myelin would be expected from the proportion of CNP⁺ and MBP⁺ cells, but also may be related to the absence of myelin-like membrane sheets in the greater proportion of CNP⁺ cells.

We also examined the expression of immunoreactive GFAP and γγ-enolase proteins, as parameters of AC cell differentiation (Fig. 5.6d, e). The increase in GFAP protein content paralleled the appearance of GFAP⁺ AC, which reached near confluency by 10 div.

Fig. 5.6 The acquisition of oligodendrocytes and the expression of phenotypic oligodendrocyte and astrocyte proteins in glial cultures maintained in serum-containing or serum-free media. Dissociated cells from neonatal mouse brain were grown in 35 mm diameter wells and maintained in DME medium containing 10% FCS (S^+ medium) for 3 days. Medium was replaced with fresh S^+ medium, or with medium consisting of DMEM/Ham's F12 (1: 1, v: v) plus supplements (S^- medium), as described in Experimental Procedures. Subsequently, S^+ and S^- media were replaced every 2 days. In each panel (a–g), values at each time point are the mean from at least 3 identical cultures ± SEM

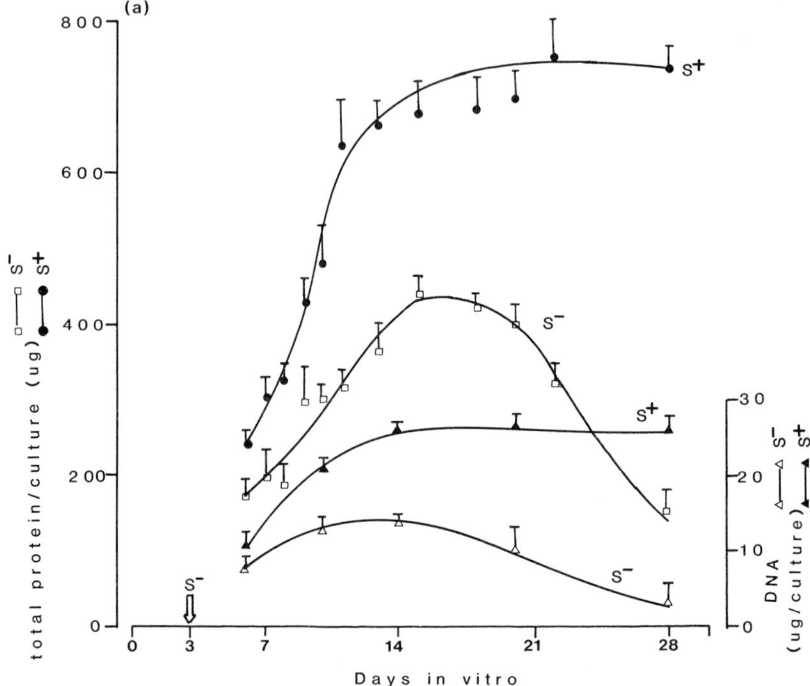

(a) Growth of cultures in S^+ and S^- media as monitored by changes in the total protein and DNA content per culture. ▲, ●, S^+ medium; △, □, S^- medium

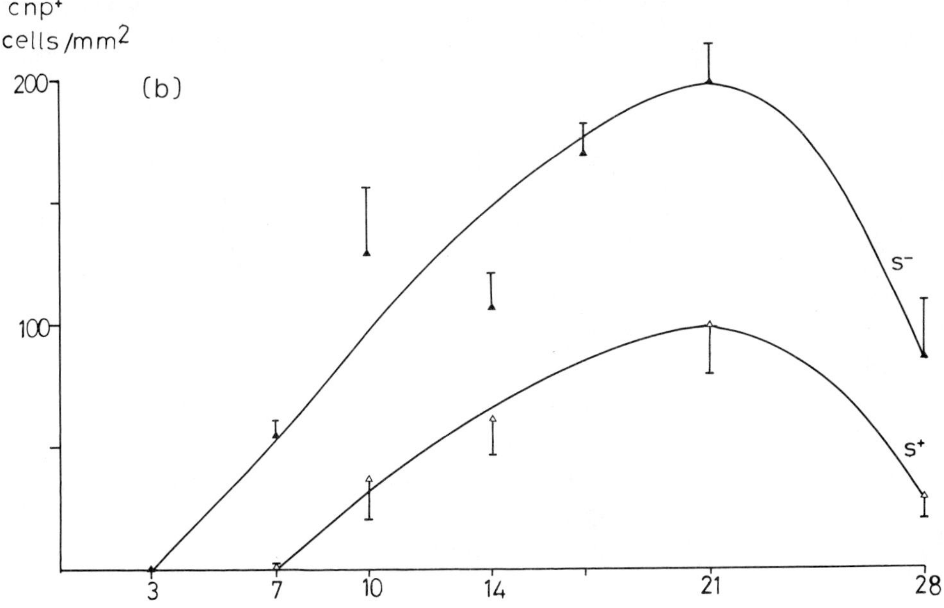

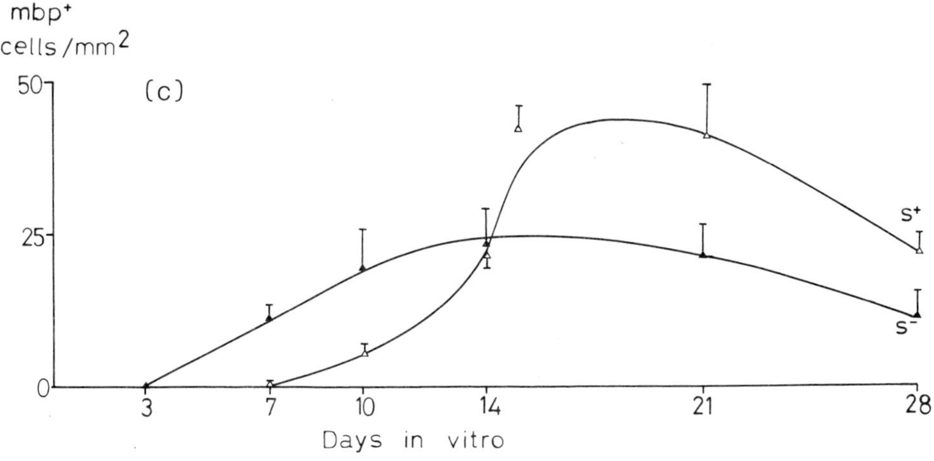

(b) and (c) The acquisition of oligodendrocytes in mixed glial cultures maintained in S^+ or S^- medium. Oligodendrocytes were detected by indirect immunofluorescence staining with rabbit polyclonal antibodies to mouse CNP (see b or to rabbit 18-kD MBP (see c and fluorescein-conjugated goat anti-rabbit IgG + IgM. Cell density was determined from average cell counts obtained from at least 10 fields of view. △——△, S^+ medium; ▲——▲, S^- medium

Fig. 5.6 continued

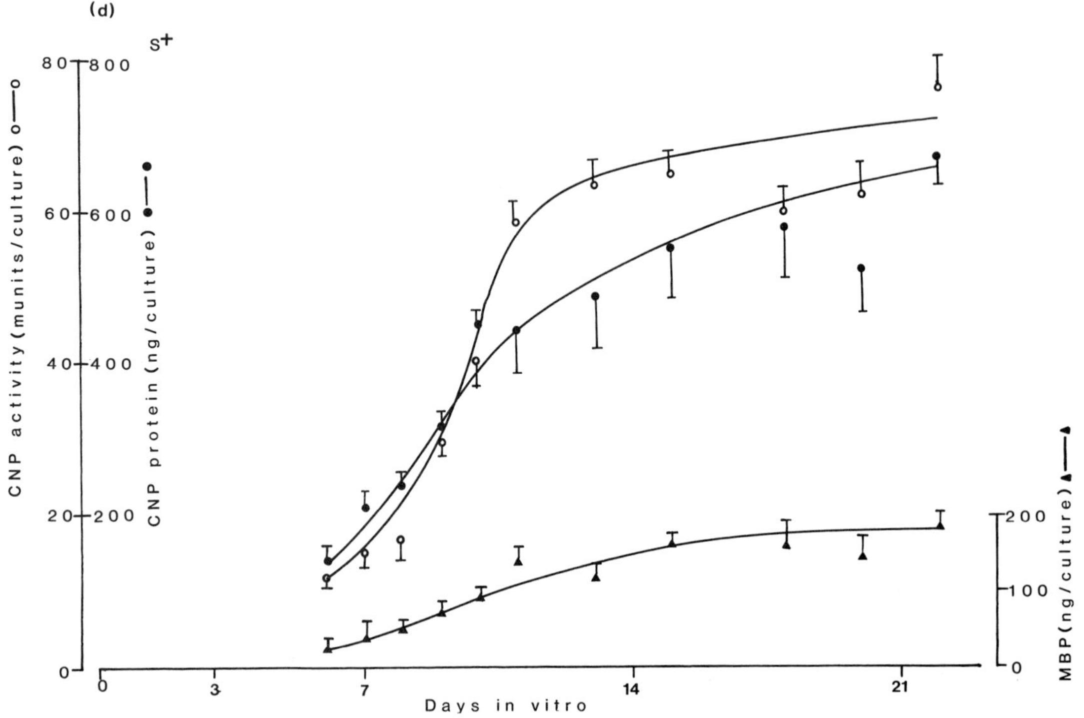

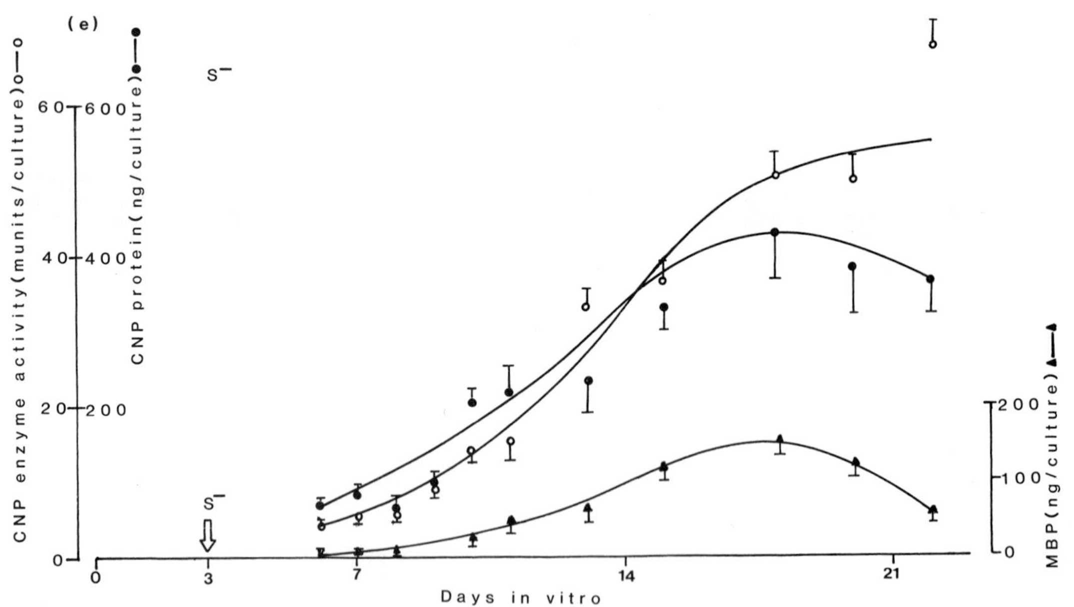

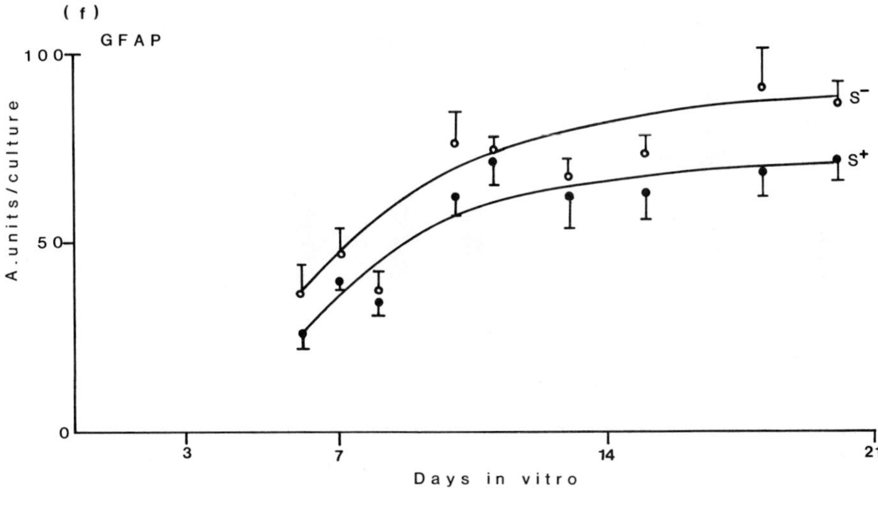

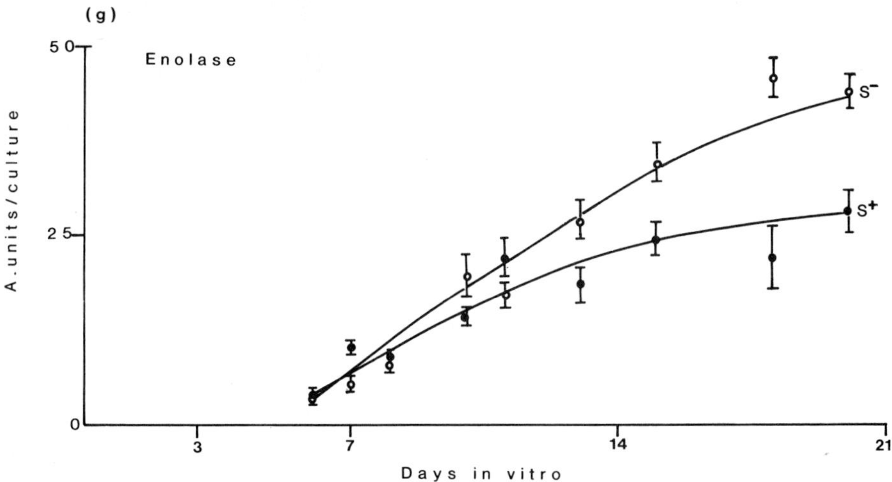

(d) and (e) The acquisition of oligodendrocyte-related proteins in mixed glial cultures. The activity of 2′, 3′ cyclic nucleotide 3′-phosphohydrolase (CNP) and the amounts of immunoreactive CNP protein, and immunoreactive myelin basic protein (MBP) was quantitated in cultures maintained in S^+ (see d) or S^- (see e) medium. Immunoreactive CNP and MBP proteins were quantitated in detergent-solubilized total cell protein by ELISA, using polyclonal antibodies and purified mouse CNP and rabbit 18-kD MBP proteins as standards. ○——○, CNP enzyme activity; ●——●, immunoreactive CNP protein; ▲——▲, immunoreactive MBP protein

(f) and (g) The acquisition of astrocyte-related proteins in mixed glial cultures. Changes in the relative amounts of glial fibrillary acidic protein (GFAP, see f) and $\gamma\gamma$-enolase (see g) were determined by ELISA. Values are expressed in terms of absorbance units obtained under the standardized conditions of the ELISA assay. ●——●, S^+ medium; ○——○, S^- medium

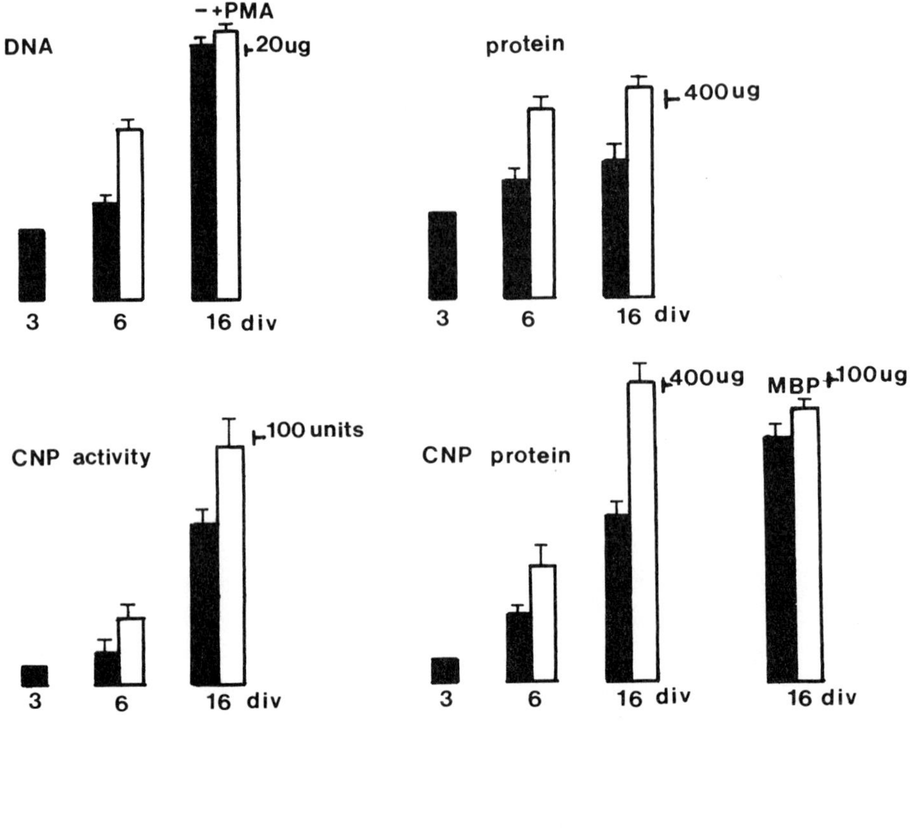

Fig. 5.7 The effect of short-term exposure to phorbol myristate acetate (PMA) on the expression of cell-type specific proteins in mixed glial cultures. Cultures in serum-free medium were exposed to 25 nM PMA for 24 h at 3 div. The content of individual immunoreactive proteins was quantitated as described in Fig. 5.6. Total DNA, protein, CNP activity and immunoreactive protein content per culture was determined at 3, 6 and 16 div.; the content of glial fibrillary acidic protein (GFAP) and $\gamma\gamma$-enolase per culture was determined at 16 div. ■, – PMA; □, + PMA at 3 div. Values are the mean ± SEM from 3 replicate cultures

Enolase protein accumulated over a longer period. Although the enolase is present in CNS neurones, the absence of neurones in mixed glial cultures from neonatal brain suggests that it is expressed by AC cells. The content of both proteins was slightly higher in S^- than S^+ medium. In contrast to ODC formation, the transfer of cultures to S^- medium had little effect on AC formation.

Because of the effect of PMA on ODC morphology, we examined the expression of CNP in PMA-treated cultures maintained in serum-free medium. Short-term exposure to PMA for 24 h at 3 div. increased the amount of CNP by 6 div. (Fig. 5.7), which fitted the impression gained by immunofluorescence staining that ODC cells had undergone hypertrophy. By 16 div., the overall number of CNP^+ cells was the same as in control cultures, although cell proliferation was accelerated and the DNA content was initially higher. There was no significant change in AC-specific proteins in response to PMA. Continuous exposure of cultures to PMA for 5 days gave rise to an initial increase in the DNA and protein content of cultures (Fig. 5.8a) and in the CNP enzyme activity (Fig. 5.8b) and immunoreactive protein content (Fig. 5.8c–f) in treated cultures over that of controls, but subsequently there was a decline. The morphological changes brought about by PMA may result in premature loss of cells through changes in cell–cell and cell–substratum contacts. Also prolonged exposure to PMA is known to render cells refractory to growth factors, which in part may be explained by loss of protein kinase C activity. This may therefore block the cellular response to insulin, preventing differentiation of precursors.

Dibutyryl-cAMP has also been reported to promote cellular differentiation. In mixed glial cultures, the addition of 1 mM dibutyryl-cAMP from 3 div. decreased cell acquisition and the protein content of cultures (Fig. 5.9a) but increased CNP protein (Fig. 5.9b). From 10 div. onwards the ratio of CNP protein to total cell protein was increased 2-fold, although the number of CNP^+ cells was reduced. The increase in CNP activity could in part be accounted for by an increase in the specific activity of the CNP enzyme protein (Fig. 5.9c), but most of the increase must be due to an increase in the average CNP content per ODC cell. It is apparent that ODC (like other mammalian cells undergoing differentiation) possess multiple stimulus–response systems for promoting cell differentiation which lead to the induction of myelin protein and membrane synthesis, with protein phosphorylation likely to play an important role in receptor–response mechanisms.

We have examined the synergistic effects of insulin and epidermal growth factor (EGF) (which has a mitogenic effect on a number of cell types) on the acquisition of ODC cells in mixed glial cultures. In the presence of insulin (10 μg/ml) or EGF (10 ng/ml) alone or together, the protein and DNA content of cultures was increased above control values (Fig. 5.10). EGF increased the DNA content to a maximum value by 15 div., whereas insulin alone gave rise to a slower increase in DNA. The addition of EGF at 3 div. gave rise to a rapid increase in the rate of AC cell proliferation, and cells exhibited a bipolar morphology. Insulin or EGF alone increased the number of CNP^+ and MBP^+ cells at 15 and 22 div. compared with cultures without either insulin or EGF (Fig. 5.11). Exposure of cells continuously to insulin in combination with EGF (added from 3–10 div. or continuously) resulted in the greatest increase (10–15 fold) in the numbers of CNP^+ and MBP^+ cells by 15 div. with no further increase at 22 div. Continuous exposure to EGF resulted in slightly fewer cells. The activity of CNP also showed the greatest increase when cultures were exposed to combinations of insulin and EGF, and the increase in CNP paralleled the increase in CNP^+ cells (Fig. 5.12). In the absence of EGF, although the total number of CNP^+ cells was low, the ratio of CNP^+ to MBP^+ cells was maximum, and most ODC expressed large myelin-like sheets. Under these conditions, at a low ODC cell density

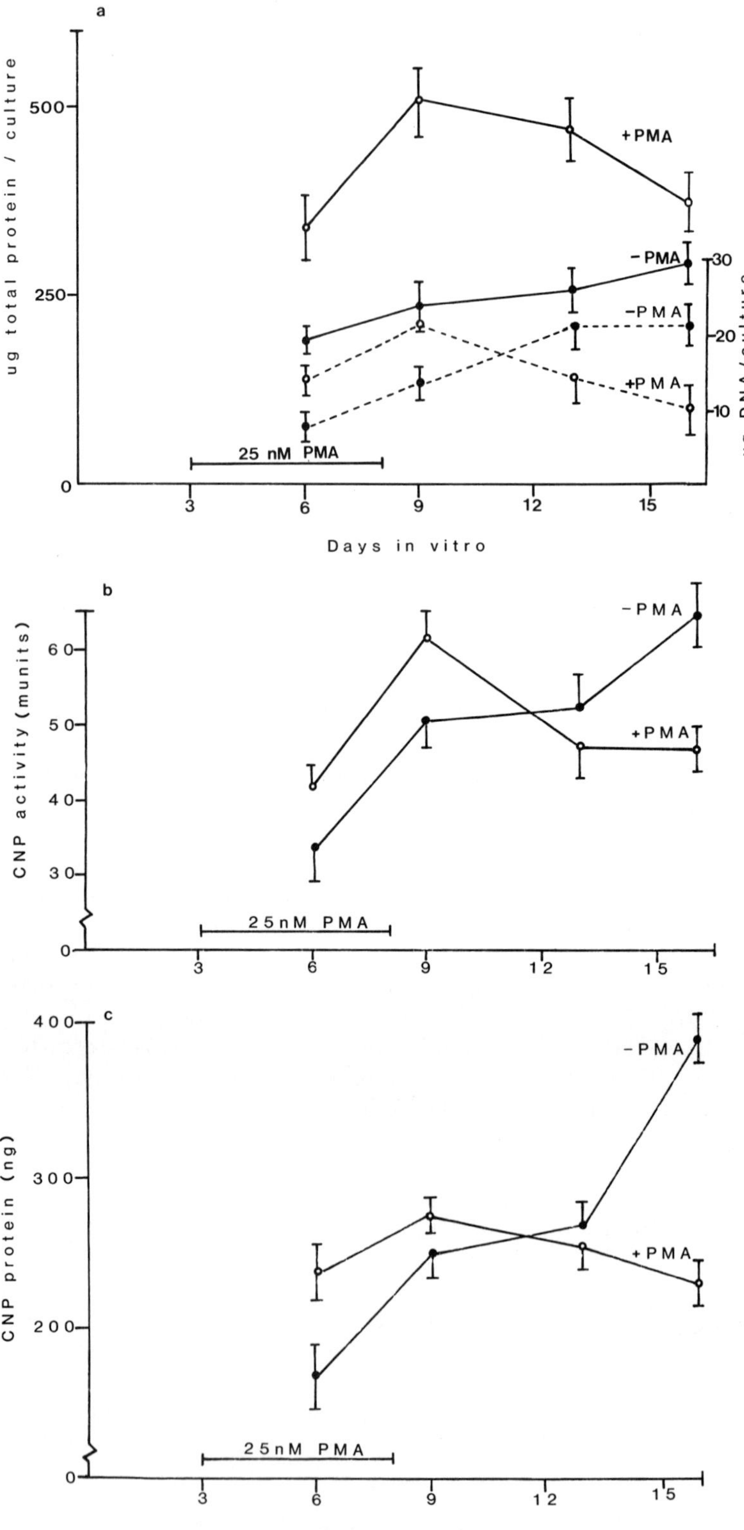
a
ug total protein / culture
500
250
0
+PMA
−PMA
−PMA
+PMA
30
20
10
ug DNA/culture
25 nM PMA
3
6
9
12
15
Days in vitro
b
CNP activity (munits)
60
50
40
30
0
−PMA
+PMA
25nM PMA
3
6
9
12
15
c
CNP protein (ng)
400
300
200
0
−PMA
+PMA
25nM PMA
3
6
9
12
15
Days in vitro

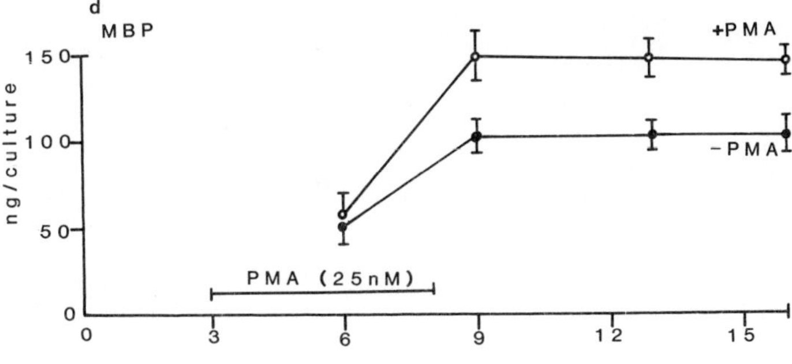

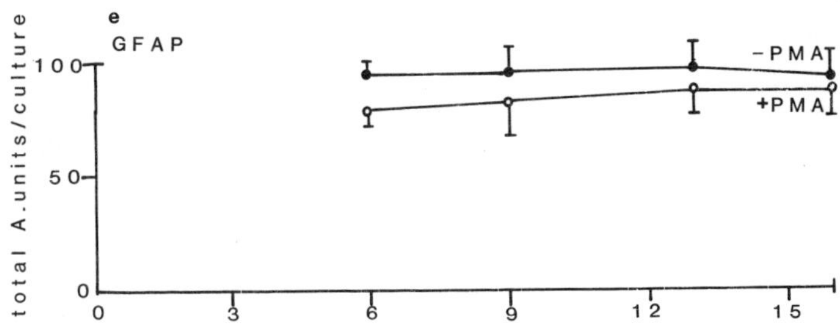

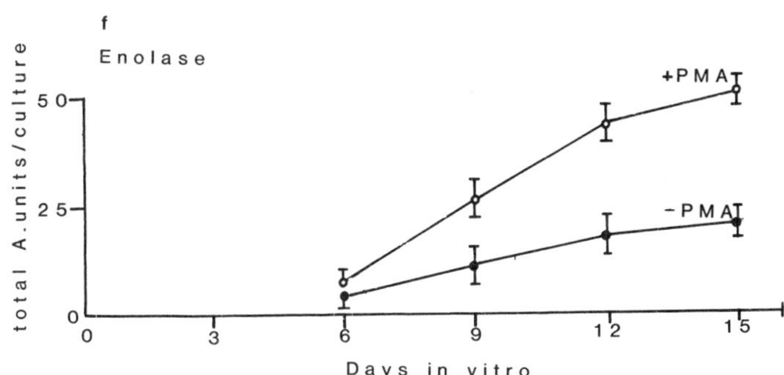

Fig. 5.8 The effect of long-term exposure to phorbol myristate acetate (PMA) on cell proliferation and the content of cell-type specific proteins in mixed glial cultures. Cultures were exposed to 25 nM PMA for 5 days from 3–8 div.: (a), total protein (———) and DNA (— — —) content per culture; (b) total CNP activity per culture; (c) total immunoreactive CNP protein per culture; (d) total immunoreactive myelin basic protein per culture; (e) total immunoreactive glial fibrillary acidic protein (GFAP) per culture; (f) total immunoreactive $\gamma\gamma$-enolase per culture. ○, +PMA; ●, −PMA. Values are the mean ± SEM from 3 replicate cultures

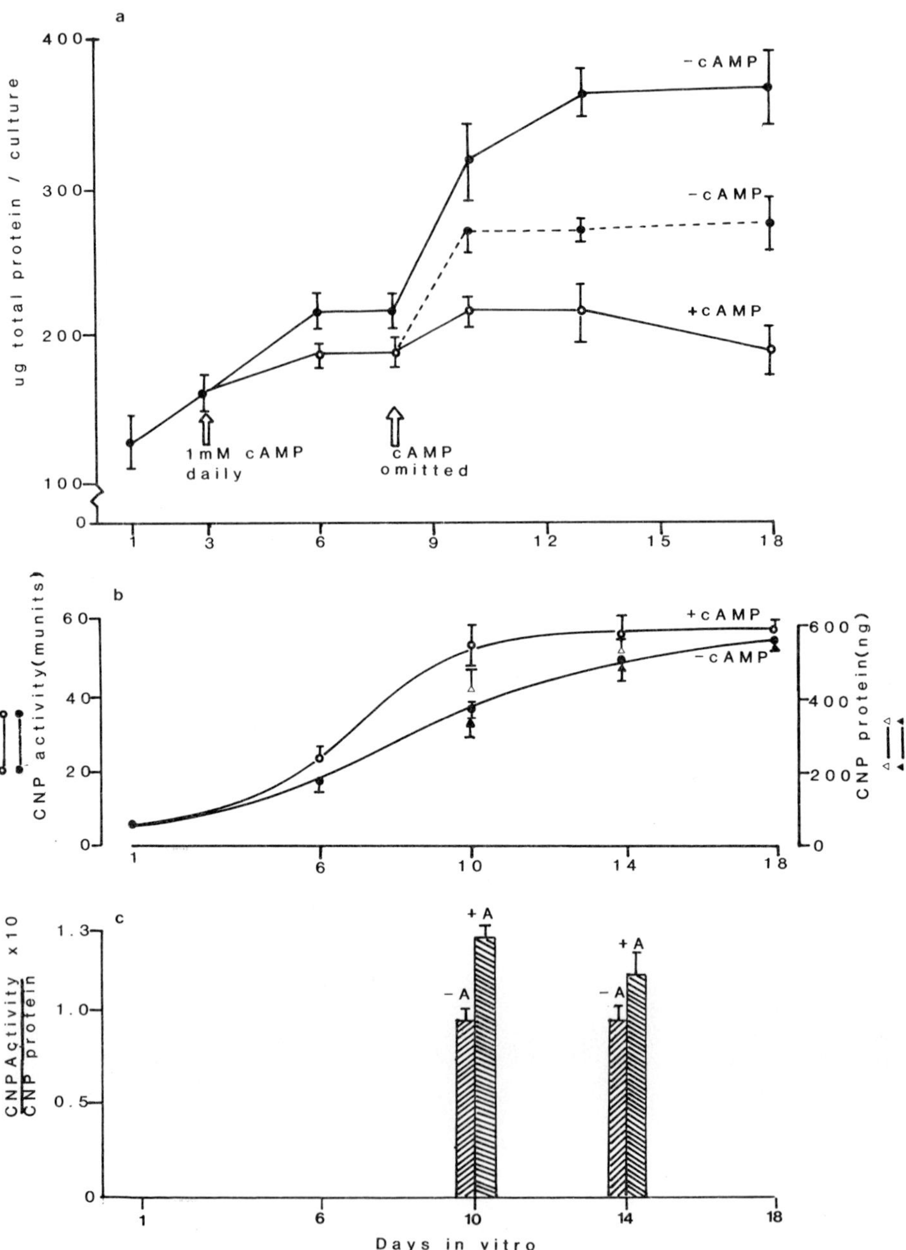

Fig. 5.9 The effect of dibutyryl 3′,5′-cyclic adenosine monophosphate on the expression of CNP in mixed glial cell cultures. 1 mM cAMP was added daily to cultures in serum-free medium as indicated, with the medium being replaced every 2 days: (a) the effect of addition of cAMP on total cell protein per culture; ○———○, +cAMP; ●———●, −cAMP; ●– – –●, cAMP omitted from 8 div.; (b) the effect of addition of cAMP on CNP enzyme activity (○———○, ●———●) and immunoreactive protein content (△———△, ▲———▲); △, ○, +cAMP; ▲, ●, −cAMP; (c) the ratio of CNP activity to immunoreactive protein in the presence (+A) or absence (−A) of cAMP. Values in (a) to (c) are the mean of triplicate cultures ± SEM. The values for +A and −A are significant ($p < 0.05$) from 6 div.

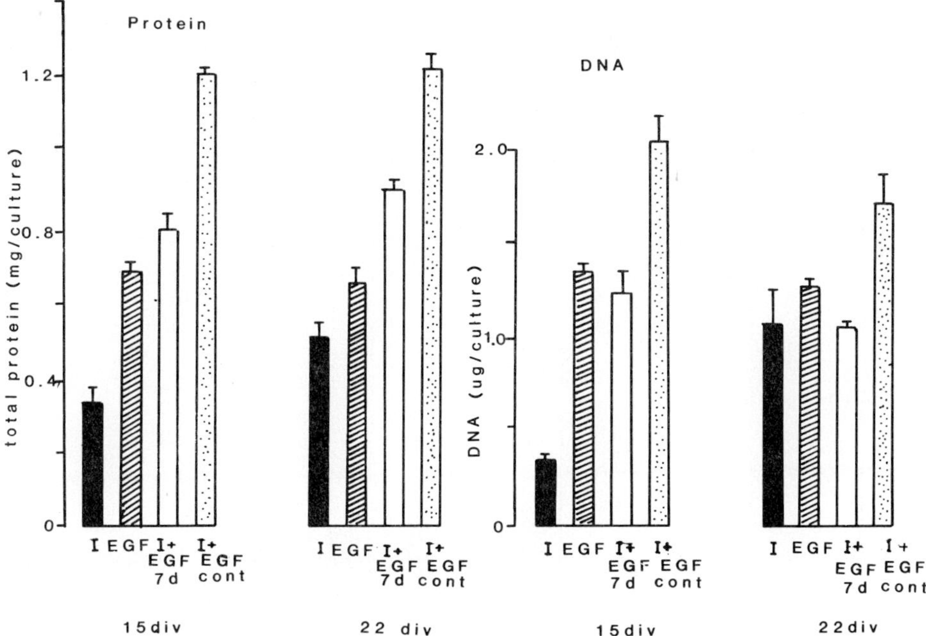

Fig. 5.10 The effect of insulin (I) and epidermal growth factor (EGF) added alone and in combination on DNA and protein acquisition in mixed glial cultures. Freshly isolated cells were maintained for 24 h in DMEM/Ham's F12 medium containing 2% UltraSer, linoleic acid/albumin and supplements (as described in Experimental Procedures) except that insulin (10 μg/ml) was present in the medium only where indicated in the figures. From 1–3 div., cultures were maintained in medium without UltraSer. At 3 div., medium was again replaced and EGF (10 ng/ml) added. Cultures were exposed to EGF for 7 days or continuously (cont) as indicated, replacing the medium every 3 or 4 days. Values are the mean of 3 replicate incubations ± SEM

and in the absence of EGF-induced mitosis, a greater proportion of cells may reach terminal differentiation promoted by insulin. Addition of EGF resulted in twice as many CNP^+ cells as MBP^+ cells, a ratio which did not change from 15–22 div., and a greater proportion of the CNP^+ ODC produced in response to EGF did not express MBP^+, and failed to reach terminal differentiation, compared with cultures in the absence of EGF. However, a lower cell density and reduced contact inhibition between ODC cells may also be a factor restricting myelin-like membrane formation (Labourdette *et al.*, 1980).

In another experiment in which insulin was present throughout, cultures were exposed to EGF for a shorter period of time (3–6 div.). At 10 div., the numbers of GC^+, CNP^+ and MBP^+ ODC were increased in response to EGF (Table 5.1). EGF also increased the number of $GFAP^+$ AC cells. Large numbers of small, round GC^+ cells proliferated in response to EGF, but few cells possessed processes. The number of GC^+ cells was already at a maximum by 10 div., and exceeded CNP^+ or MBP^+ cells at both 10 and 17 div. Cytosine arabinoside, which is an anti-mitotic agent, blocked the acquisition of both ODC and AC when added at 3–6 div., but had no effect when added at 6–10 div. There was no further increase in GC^+ or CNP^+ ODC on addition of EGF at 10–13 div. to cultures previously

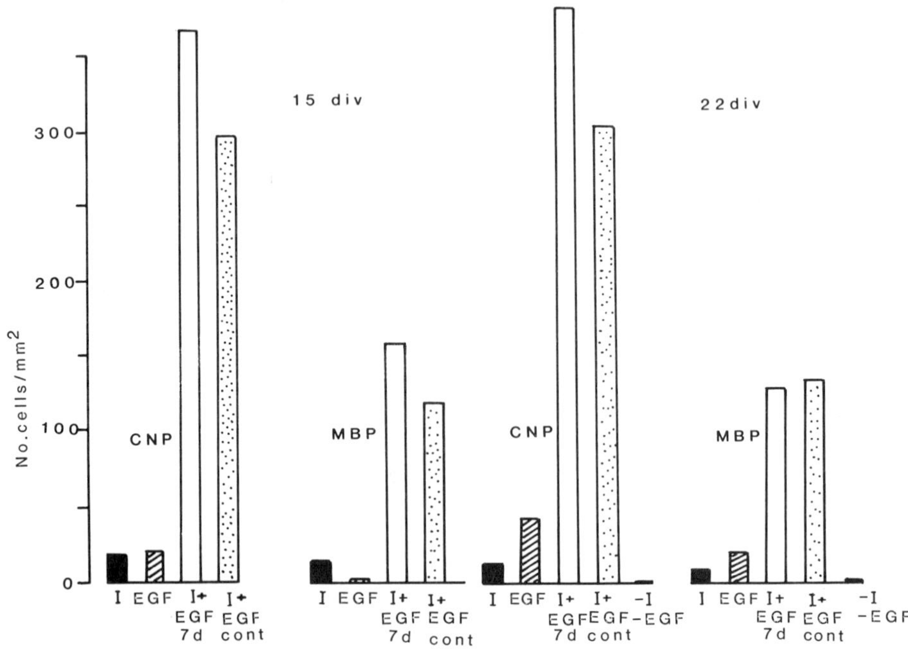

Fig. 5.11 The effect of insulin (I) and epidermal growth factor (EGF) on the acquisition of oligodendrocytes in mixed glial cultures. Data were obtained from the same cultures as in Fig. 5.10.

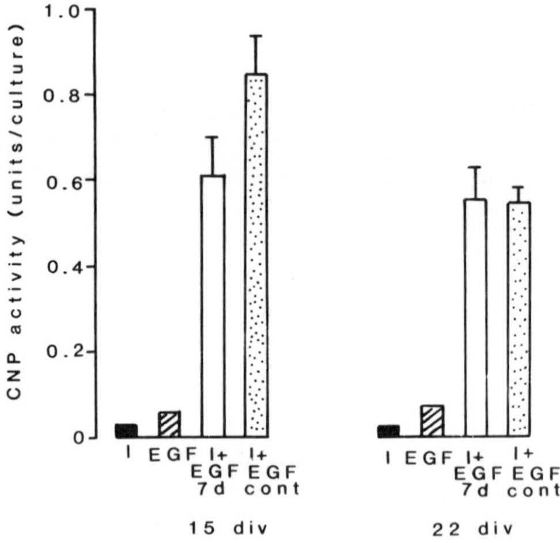

Fig. 5.12 The effect of insulin (I) and epidermal growth factor (EGF) on the expression of CNP activity in mixed glial cultures. Data were obtained from the same cultures as in Figs. 5.10 and 5.11

exposed to either EGF or cytosine arabinoside, and the results indicate that the EGF stimulation of ODC precursor proliferation is maximal around 3–6 div. Most of the GC$^+$ ODC precursors, which were present by 10 div., did not differentiate further in the presence of insulin with or without EGF. There was no evidence for a pool of ODC precursor cells in culture after 10 div., which are responsive to the mitotic stimulus of EGF, or the differentiation stimulus of insulin.

Mercanti *et al.* (1987) also found that placental mitogens stimulate the proliferation of ODC precursors, which were present by 10 div., did not differentiate further in the presence the present author, McMorris *et al.* (1984) found that insulin increased the number of ODC cells in mixed glial cultures. However, there is little evidence for insulin crossing the blood–brain barrier and the concentrations of insulin used to elicit responses in culture are non-physiological. It seems likely that insulin mediates its effect by binding to insulin-like growth factor (IGF) receptors. If this is so, then IGF peptides may play an important role in ODC differentiation. The synergism between EGF and insulin indicates that both EGF-like and insulin-like peptides are required, acting through separate effector systems as in other cell types (Rozengurt, 1986). The timing of the mitotic response to EGF implies that the receptor–response system for EGF is lost by 10 div., and this may allow non-dividing cells to differentiate in response to insulin receptor activation.

The effect of cell–cell contacts and, in culture, cell–substratum contacts may also act to stimulate or inhibit ODC differentiation, depending on the type of interaction (Labourdette *et al.*, 1980; Vartanian *et al.*, 1986). In culture, a low concentration of ODC appears to favour terminal differentiation, and ODC–ODC cell contacts may inhibit the formation of myelin-like membrane sheets.

Summary

It has been shown that glial cell cultures provide a useful model system to study ODC cell formation, myelinogenesis and the cellular response mechanisms to specific growth factors. Investigation of the synergistic effects of EGF-like mitogenic growth factor peptides and insulin-like growth factors on ODC differentiation and myelinogenesis in culture contribute to a better understanding of the process of myelination in the infant brain, hormonal- and growth factor-related defects in infant brain development, and provide information on the potential for the induction of re-myelination in the adult brain.

Acknowledgements

Part of this work was supported by a grant from the Swiss National Science Fund, and is currently being supported by the UK Medical Research Council. The financial assistance of The Humane Research Trust in the purchase of essential equipment for cell culture is gratefully acknowledged.

References

Bologa, S. L., Siegrist, H. P., Z'graggen, A., Hofmann, K., Wiesmann, U., Dahl, D. and Herschkowitz, N. (1981). *Brain Res.*, **210**, 217–29.

Blake, M. S., Johnson, K. D., Russel-Jones, G. J. and Gotschlich, E. C. (1984). *Anal. Biochem.*, **136**, 175–9.

Carey, E. M. and Foster, P. C. (1984). *J. Neurochem.*, **42**, 924–9.
Ecclestone, P. A. and Silberberg, D. B. (1984). *Dev. Brain Res.*, **16**, 1–9.
Foster, P. C. and Carey, E. M. (1983). *Early hum. Dev.*, **9**, 33–47.
Kim, S. U., McMorris, F. A. and Sprinkle, T. J. (1984). *Brain Res.*, **300**, 195–9.
Labourdette, G., Russel, G. and Nussbaum, J. L. (1980). *Neurosci. Lett.*, **18**, 203–9.
McCarthy, K. D. and de Vellis, J. (1980). *J. Cell Biol.*, **85**, 890–902.
McMorris, F., Kim, S. U. and Sprinkle, T. J. (1984). *Brain Res.*, **292**, 123–31.
Mercanti, D., Luzzatto, E., Ciotti, M. J. and Levi, G. (1987). *Expl Cell Res.*, **168**, 182–90.
Merril, C. R., Dunan, M. L. and Goldman, D. (1981). *Anal. Biochem.*, **110**, 201–7.
Mimms, L. J., Zampigni, G., Nozaki, Y., Tanford, C. and Reynolds, J. A. (1981). *Biochemistry (USA)*, **20**, 833–40.
Reynolds, R., Carey, E. M. and Herschkowitz, N. (1988). *Neuroscience*, in press.
Raff, M. C., Miller, M. and Noble, M. (1983a). *Nature*, **303**, 390–6.
Raff, M. C., Abney, E. A., Cohen, J., Lindsay, R. and Noble, M. (1983b). *J. Neurol. Sci.*, **3**, 1289.
Roussel, G., Sensenbrenner, M. and Labourdette, G., Wittendorp-Rechenmann, E., Pettman, B. and Nussbaum, J. L. (1983). *Dev. Brain Res.*, **8**, 193–204.
Rozengurt, E. (1986). *Science*, **234**, 161–6.
Sprinkle, T. J., Sheedlo, H. J., Buxton, T. B. and Rissing, J. P. (1983). *J. Neurochem.*, **41**, 1664–71.
Szuchet, S. and Stefansson, K. (1980). *Adv. Cell Neurobiol.*, **1**, 313–46.
Vartanian, T., Szuchet, S., Dawson, G. and Campagnoni, A. T. (1986) *Science*, **234**, 1395–8.

CHAPTER 6

The use of tumour cell cultures for understanding neuroendocrine cell biology

J. M. Polak and S. R. Bloom

Introduction

The intracellular events which take place during peptide hormone synthesis and release by a neuroendocrine tumour cell are still poorly understood. One way of investigating this problem is by developing an animal model. However, in spite of intense efforts, this has been rather fruitless, with the exception of a few experimentally-induced islet cell tumours of the pancreas including those produced by novel gene transfer technology (Hanahan, 1985).

In contrast to animal experiments, where there is inherent variability between animals, tumour cell lines are of considerable advantage in experimental work, since it is possible to produce large numbers of identical cells in exactly defined and maintained conditions.

The diffuse neuroendocrine system and its derivative tumours

Since Feyrter's original description (Feyrter, 1953) of the existence of specialized endocrine cells dispersed throughout the body, the concept of the 'diffuse endocrine system' has expanded in order to include a special component of the nervous system which, like the specialized endocrine cells, is capable of producing and releasing active peptides; hence the name 'diffuse neuroendocrine system' (Polak and Bloom, 1979). What is a regulatory peptide? The term was introduced (Polak and Bloom, 1983) when it was realized that active peptides are produced by specialized sets of endocrine cells and nerves and released either via the circulation, acting at a distance, or locally, acting as 'paracrine' (local) hormones (see later) or neurotransmitters. In general, the peptide-producing endocrine cells stain rather poorly with conventional histological methods, and were thus also called 'clear' cells. A number of specialized techniques have, however, been in use for some time (Figs 6.1–6.3). These include a variety of silver impregnation methods (Grimelius and Wilander, 1985), in particular that of Grimelius, and the use of a battery of antibodies to the

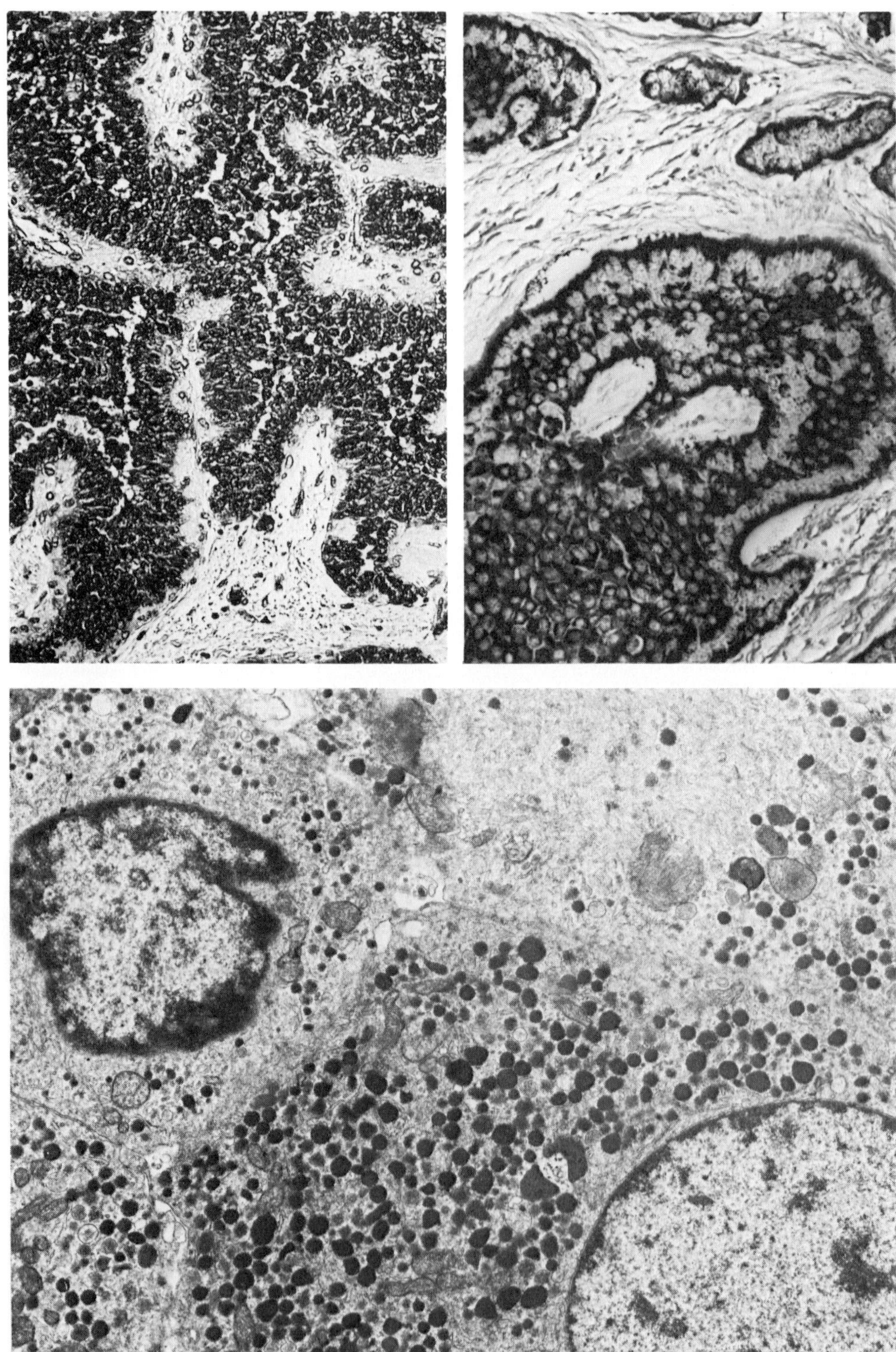

so-called 'general neuroendocrine markers', including neuron-specific enolase (Polak and Marangos, 1984), an isomer of the glycolytic enzyme enolase, and chromogranins (Wilson and Lloyd, 1984), a family of proteins originally extracted from the adrenal medulla, and associated with secretory granules. Electron microscopy is particularly valuable for demonstrating neurosecretory granules which, fortunately, survive well, even in poorly fixed material.

The tumours arising in the diffuse neuroendocrine system are variously termed Apudomas (the cells and nerves of the diffuse neuroendocrine system are also recognized as APUD cells (*A*mine *P*recursor *U*ptake and *D*ecarboxylation)), a term introduced by Pearse (1969), neuroendocrine tumours, carcinoid tumours, islet cell tumours (if arising in pancreatic islets) or medullary carcinomas (a tumour of thyroid C cells) (Polak and Bloom, 1985). Like normal neuroendocrine cells, neuroendocrine tumours contain dense-cored secretory granules and can be stained using silver impregnation methods and by immunocytochemistry using either antibodies to the so called 'general neuroendocrine markers' or to specific peptides for the demonstration of hormonal activity. The production of active hormones by neuroendocrine tumours has been increasingly recognized of late, and classical clinical features due to the excessive production of peptide hormones by tumours have been frequently described (Ch'ng *et al.*, 1985). Furthermore, the proposed growth-promoting properties of certain peptide hormones have been further elucidated by the analysis of particular classes of neuroendocrine tumours. An example of this is the characteristic neuroendocrine neoplasm of the lung termed 'small cell carcinoma' (Cuttitta *et al.*, 1985). This neoplasm, which is the second most common tumour of the lung, is often associated with the production and release of a regulatory peptide termed bombesin. Mammalian bombesin or gastrin-releasing peptide (GRP) is a 27 amino acid peptide with close chemical similarities to the bombesin originally discovered in amphibia. Increasing circumstantial evidence indicates that growth of a small cell carcinoma of the lung (SCCL), depends in part on the production, release and re-binding to a specific receptor, of mammalian bombesin. Thus, mammalian bombesin may be a growth-promoting factor responsible for the growth of a small cell carcinoma of the lung (Cuttitta *et al.*, 1985). Evidence for this includes the finding of elevated levels of mammalian bombesin in a large proportion of small cell carcinomas grown in culture, the observation that small cell carcinoma cell lines grow faster if pure bombesin is added to the medium and the discovery that this growth is retarded after the addition of specific monoclonal antibodies to bombesin. Furthermore, specific binding sites for bombesin have been identified on the surface of these tumour cells, both biochemically and morphologically (Lackie *et al.*, 1985). It is interesting that the structure of human probombesin has recently been elucidated by using techniques of molecular biology from a neuroendocrine tumour of the lung (Spindel *et al.*, 1984). The molecule is composed of 266 amino acids and in the human prohormone bombesin is flanked by amino acid sequences at both the N- and C-terminals. Furthermore, it has been shown that mammalian bombesin originates from three separate

Fig. 6.1 (top left) NSE-immunoreactive cells in a neuroendocrine tumour. Formalin-fixed, 5μm wax section, peroxidase antiperoxidase method. (× 100)

Fig. 6.2 (top right) Grimelius' siver impregnation method in a neuroendocrine tumour. Formalin-fixed, 5μm wax section. (× 250)

Fig. 6.3 (bottom) Neuroendocrine tumour of the pancreas. Two types of dense-core secretory granules are seen in two separate cell types. (× 12000)

precursors, the product of three mRNAs giving rise to three different C-terminal flanking peptides of variable lengths (Sauville *et al.*, 1986; Spindel *et al.*, 1986) (see Fig. 6.4).

The mechanisms by which peptides with growth-promoting properties are produced and transported within the cell and finally secreted, are of interest, as are the ways in which their trophic actions are mediated by receptors (Rozengurt, 1986). We therefore focused our studies on the biology of the small cell carcinoma of the lung.

Specimens used and technology employed

SURGICAL SAMPLES OF LUNG NEUROENDOCRINE TUMOURS

Formalin-fixed samples were obtained from the files of the Departments of Histopathology at the Hammersmith Hospital (London), the Brompton Hospital (London) and Frenchay Hospital (Bristol). When new cases were available, blocks were fixed in Bouin's solution or buffered formalin and, for electron microscopy, in 2.5% glutaraldehyde for 2 h.

SMALL CELL CARCINOMA CELL LINES

Cell lines were obtained from the National Cancer Institute (Washington, USA) (Oie *et al.*, 1984) and one cell line was raised in the Department of Histochemistry, Hammersmith Hospital. Two types of media were used: serum-supplemented media, containing a variety of undefined growth factors, and serum-free media with defined medium composition (Oie *et al.*, 1984). Cells were maintained in serum-supplemented media (RPMI 1640 + 10% foetal calf serum). This was also used to ensure that the response of the cells was not

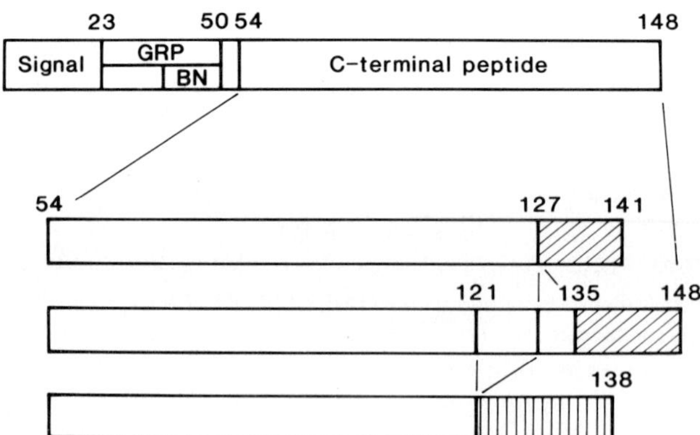

Fig. 6.4 Diagrammatic representation of the precursor of human bombesin as predicted from analysis of the gene. The whole molecule is shown at the top, and the three sequences of the C-terminal flanking peptide are enlarged below. These three peptides have an identical N-terminal portion but variable C-terminal sequences (shaded). Two of the molecules are very similar except that 7 amino acids are deleted in the C-terminal end. The third has a 19-base deletion in the mRNA coding sequence, which, in addition to deleting 7 amino acids also results in a shift in the reading frame and therefore a different C-terminal amino acid sequence from the other two

restricted in some unexpected way by serum-free media. Cells for most experiments were grown in serum-free media. Small cell carcinoma cells were grown in flasks under sterile conditions, floating in cell culture medium in a humidity-controlled atmosphere containing 5% ± 0.1% CO_2 (which is required for cell growth and additionally helps to maintain the pH of the medium). A constant temperature of 37°C ± 0.1°C was maintained.

IMMUNOCYTOCHEMISTRY

Demonstration of neuroendocrine differentiation and peptide production was carried out by immunocytochemistry (Polak and Van Noorden, 1986) using antibodies to neuron-specific enolase, to bombesin, and to the first 21 amino acids of the C-terminal peptide of human pro-bombesin.

RECEPTOR LOCALIZATION.

The techniques commonly used for receptor localization, viz. *in vitro* autoradiography (Young and Kuhar, 1979) and the use of antibodies to the receptor molecule (Skovgaard Poulson *et al.*, 1985) were not applicable in this particular study. The former, because of poor resolution for electron microscopy, and the latter, because pure preparations of receptors for immunization procedures were not available. Instead, we used a novel procedure which is based on the use of a monoclonal antibody to mammalian bombesin of known region-specificity. In this particular case, a monoclonal antibody recognizing the active C-terminal portion of bombesin known to attach to specific bombesin receptors was used (Lackie *et al.*, 1985). It was therefore necessary to construct a divalent form of bombesin having two reactive C-terminal sites—one that would attach to the receptor and another that would be free to recognize the bombesin monoclonal antibody (Fig. 6.5) (Lackie *et al.* 1985). The final reaction product was visualised at the electron microscopical level by gold-labelling procedures (Polak and Varndell, 1984). Ultrathin frozen sections were employed in order to avoid denaturing receptor and antigenic sites.

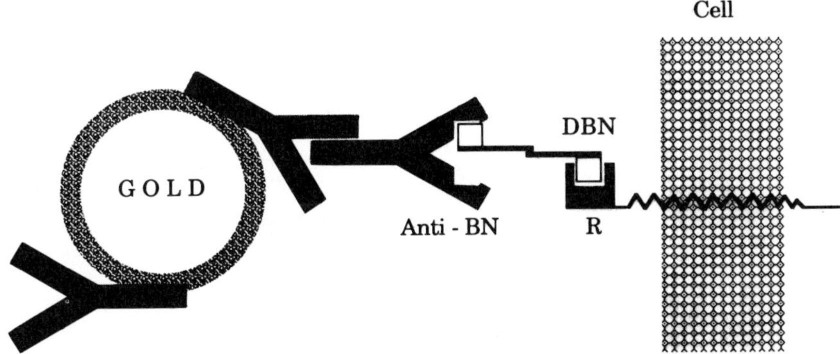

Fig. 6.5 Diagrammatic representation of the binding of a divalent bombesin ligand (DBN) to a cell surface receptor (R) and its visualization with immunogold-labelled antibodies

MANIPULATION OF CELL LINES

In order to disrupt the process of synthesis, packaging, secretion and receptor recognition, cultures were manipulated in several ways (Lackie, 1987). They were treated with pharmacological agents to enhance granulation and suppress secretion: with microtubule- and microfilament-dependent transport and secretion, blockers included colchicine, cytochalasin B, cytochalasin D, nocodazole and tubuzole. This last is a synthetic stereospecific blocker; its inactive stereo-isomer was also used as a pharmacological control. Alterations in the culture regime and frequent changing of the culture medium were also used for the same purpose. In order to establish highly dense or loosely-packed cell lines different nutrients and different oxygen concentrations were used. Theophylline, on the other hand, was used to cause cell degranulation by maintaining the levels of cAMP in tumour cells.

IN SITU HYBRIDOHISTOCHEMISTRY

Peptide immunostaining and measurement analyses only stored peptide. In order to understand tumour cell biology it is necessary to employ other more 'functional' techniques, e.g. analysis of peptide binding sites by receptor localization. Another possibility is to localize the mRNA that directs peptide synthesis. This is now possible by the use of a technique called *in situ* hybridohistochemistry (Hudson *et al.*, 1981). This technique makes use of available specific DNA sequences which can hybridize with specific intracellular mRNA. Initially, complementary DNA sequences (single or double stranded) were the probes most commonly used but lately the use of complementary RNA sequences is the technique of choice. The probes can be radioactively or non-radioactively labelled and can be analysed at light or electron microscopical level. We have been able to use radiolabelled cRNA probes recognizing the mRNA of mammalian bombesin (Fig. 6.6).

Staining of surgical samples

We have been able to carry out extensive retrospective and prospective immunocytochemical analysis on a wide range of pulmonary and non-pulmonary neuroendocrine tumours, using antibodies to a general neuroendocrine marker (i.e. neuron-specific enolase) and to both bombesin and the C-terminal flanking peptide of human probombesin (Hamid *et al.*, 1987). These results are summarized in Table 6.1 and Fig. 6.7.

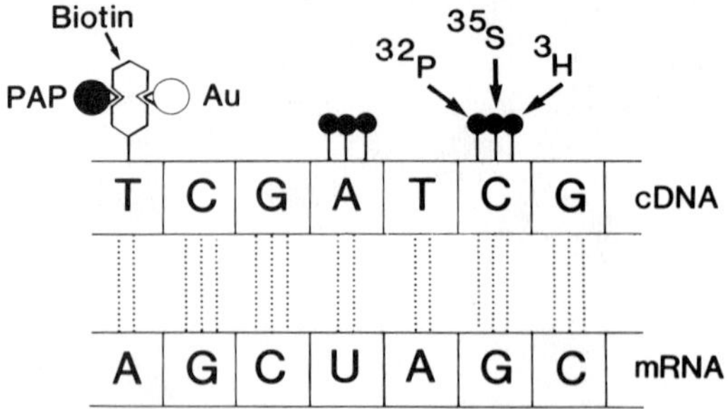

Fig. 6.6 Schematic representation of the *in situ* hybridization method using a cDNA probe, carrying different possible labelling methods necessary to visualize the reaction product

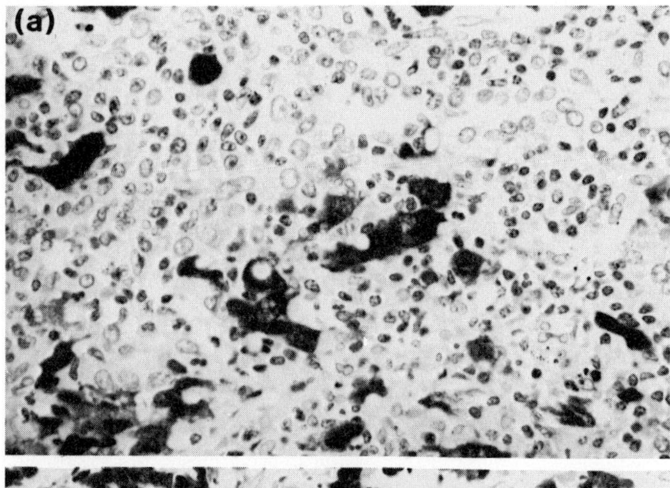

Fig. 6.7 (a) Small cell carcinoma of the lung fixed in formalin and immunostained using bombesin antibodies. Peroxidase antiperoxidase. Note only scattered cells are stained. (× 500)

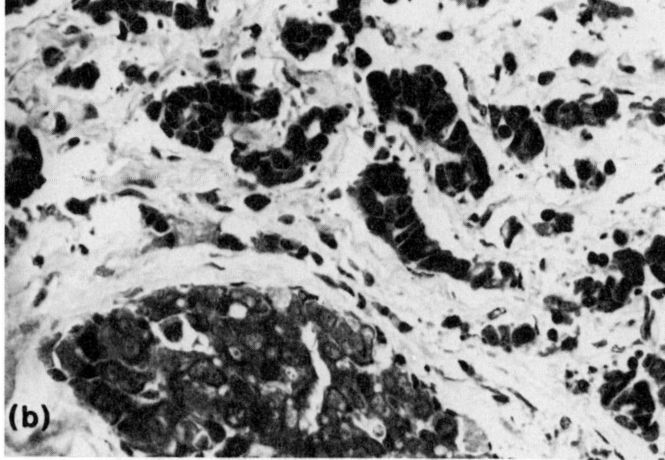

(b) Atypical (aggresive) carcinoid of the lung immunostained with an antibody to the C-terminal flanking peptide of human bombesin. Formalin fixation. Peroxidase antiperoxidase staining. Note intense and abundant staining of tumour cells. (× 500)

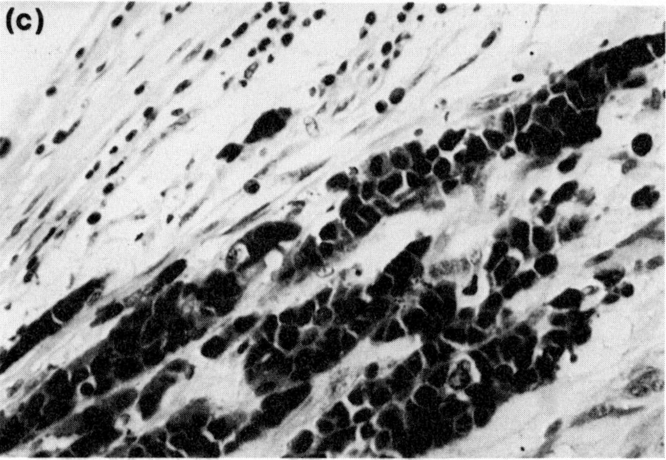

(c) Small cell carcinoma of the lung fixed in formalin and immunostained with antibodies to the C-terminal flanking peptide of human bombesin by the peroxidase antiperoxidase method. Note intense staining of most tumour cells. (× 500)

Table 6.1 Number of cases and percentage of positive results using antibodies to general neuroendocrine markers and to bombesin gene products in 353 tumours of the lung

		*Antibody**			
Tumour	*Total*	*NSE*	*PGP*	*BN*	*CFP*
Small cell carcinoma (oat cell + intermediate)	156	122 (78%)	141 (90%)	38 (24%)	104 (67%)
Combined small cell carcinoma	5	4 (80%)	5 (100%)	1 (20%)	4 (80%)
Atypical carcinoid	32	22 (69%)	25 (78%)	9 (28%)	19 (56%)
Carcinoid	51	44 (86%)	47 (97%)	17 (33%)	6 (12%)
Non-small cell carcinoma: squamous cell (46), adenocarcinoma (39), large cell undifferentiated (24)	109	13 (11%)	40 (35%)	0 (0%)	0 (0%)
Total	353				

**Key*: NSE, neuron specific enolase; PGP, protein gene product; BN, bombesin; CFP, C-terminal flanking peptide of human bombesin.

Tumour cells in culture

Two distinct patterns of growth were seen in the suspension cultures. The slow growing one showed the 'hollow sphere' pattern of growth, as described for 'classical' line, whereas the so-called 'variant' cell line grew as loose aggregates (Fig. 6.8).

The cell lines displayed the classical characteristics of neuroendocrine tumour cells in cultures. Many of the cells produced long, thin processes often containing electron dense-cored granules (80–120 nm in diameter) and showed a high nucleus to cytoplasm ratio, although granules were sparse, in particular in the 'variant' type. Cells in culture showed a staining pattern similar to that seen in surgical samples, including immunoreactivity for neuron-specific enolase and bombesin. Receptors were visualized on the surface of the cultured tumour cells (Fig. 6.9) and mRNA-directing bombesin production was visualized by *in situ* hybridohistochemistry methods within the cells. Changes after culture manipulations can be seen in Table 6.2.

Summary

Neuroendocrine tumours present interesting problems in cell biology and those of the lung, in particular, provide an ideal model for investigation of the biology of peptide-producing cells. Unfortunately, surgical samples are unsuitable for studies involving manipulation of

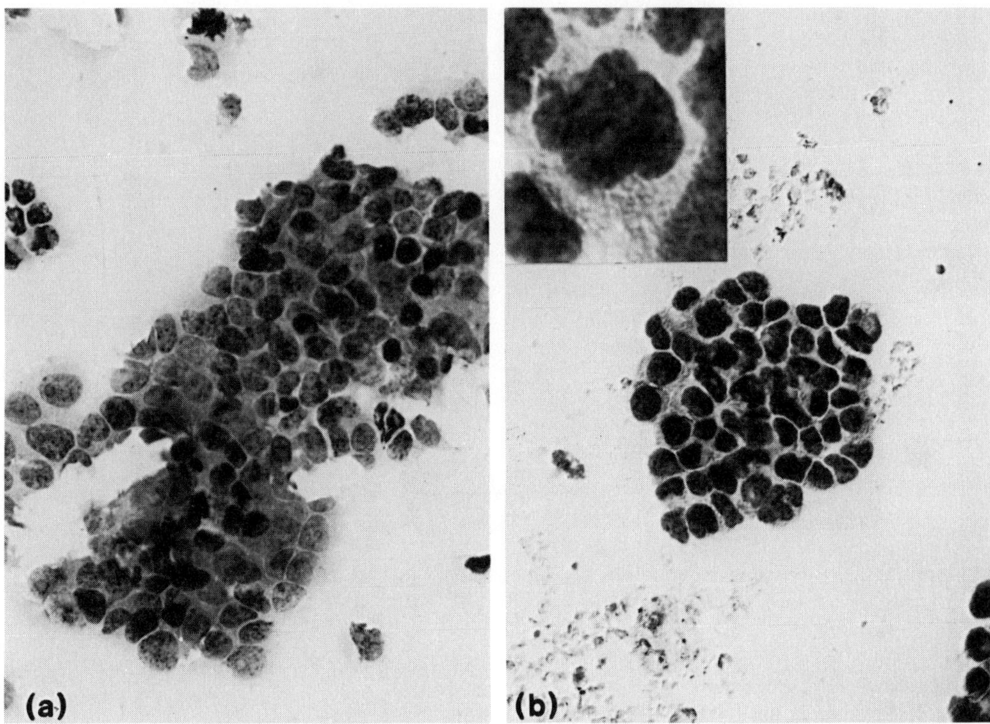

Fig. 6.8 (a) Conventional (haematoxylin–eosin) staining of small cell carcinoma cells in culture; 4% paraformaldehyde fixation, cytospin preparation. (× 500)

(b) Same cell line immunostained using an antibody to the C-terminal flanking peptide of bombesin cells. The inset shows the details of an immunoreactive cells; 4% paraformaldehyde fixation, cytospin preparation. Peroxidase antiperoxidase method. (× 500; inset × 2000)

Table 6.2 Culture manipulations*

Agent	*Effect*	*Outcome*
Colchicine Tubulozone	Microtubule blocker	Granulation +
Cytocholasin B Cytocholasin D	Microfilament blocker	Granulation ± or −
Theophylline	[cAMP] +	Granulation −
Frequent changes of medium	?	Granulation +

* +, increased; −, decreased; ±, equivocal.

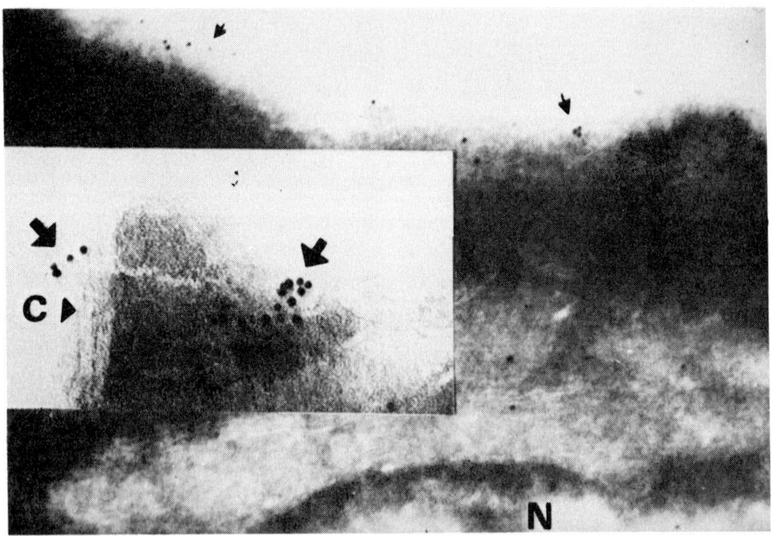

Fig. 6.9 Ultrastructural visualization of bombesin receptors (arrows) on the surface of a cultured small cell carcinoma using gold-labelled antibodies. C = cell membrane, N = nucleus. (× 180 000)

mRNA or hormone expression and experimentally-induced tumours in animals are undesirable and difficult to produce. Tumour cell lines thus provide an excellent model. The use of cell lines and the application of modern microscopical imaging techniques combined with biochemical analysis allows a growing understanding of neuroendocrine cell biology.

Acknowledgements

The generous support of the Humane Research Trust, the Cancer Research Campaign and the Council for Tobacco Research USA, is acknowledged.

References

Ch'ng, J. L. C., Polak, J. M. and Bloom, S. R. (1985). In *Endocrine Tumours. Current Problems in Tumour Pathology*. Eds Polak, J. M. and Bloom, S. R., pp. 264–80. Edinburgh, Churchill Livingstone.

Cuttitta, F., Carney, D. N., Mulshine, J., Moody, T. W., Fedorko, J., Fischler, A. and Minna, J. D. (1985). Bombesin-like peptide can function as autocrine growth factor in human small cell cancer. *Nature*, **316**, 823–6.

Feyrter, F. (1953). Uber die peripheren endokrinen (parakrinen) Drüsen des Menschen. Wien, Maudrich.

Grimelius, L. and Wilander, E. (1985). In *Endocrine Tumours. Current Problems in Tumour Pathology*, Eds Polak, J. M. and Bloom, S. R., pp. 95–115. Edinburgh, Churchill Livingstone.

Hamid, Q., Addis, B. J., Springall, D. R., Ibrahim, N. B. N., Ghatei, M. A., Bloom, S. R. and Polak, J. M. (1987). Expression of C-terminal peptide of human pro-bombesin in 361 lung endocrine tumours, a reliable marker and possible prognostic indicator for small cell carcinoma. *Virchows Arch. Pathol.*, **411**, 185–92.

Hanahan, D. (1985). Heritable formation of pancreatic B-cell tumours in transgenic mice expressing recombinant insulin/Simian virus 40 oncogenes. *Nature*, **315**, 115–22.

Hudson, P., Penschow, J., Shine, J., Ryan, G., Niall, H. and Coghlan, J. (1981). 'Hybridisation histochemitry': Use of recombinant DNA as a 'homing probe' for tissue localisation of specific mRNA populations. *Endocrinology*, **108**, 353–6.

Lackie, P. M. (1987). Peptide Containing Endocrine Cells and Endocrine Tumours in the Respiratory Tract; Investigated by Cell Culture Immunocytochemistry and Electron Microscopy. PhD thesis, University of London.

Lackie, P. M., Cuttitta, F., Minna, J. D., Bloom, S. R. and Polak, J. M. (1985). Localisation of receptors using a dimeric ligand and electron immunocytochemistry. *Histochemistry*, **83**, 57–9.

Oie, H. K. K., Brower, M. and Carney, D. N. (1984). Growth factor requirements for *in vitro* growth of endocrine and nonendocrine lung cancers in serum-free defined media. In *The Endocrine Lung in Health and Disease*. Eds Becker, K. L. and Gazdar, A. F., pp. 469–75. Philadelphia, W. B. Saunders.

Pearse, A. G. E. (1969). The cytochemistry and ultrastructure of polypeptide hormone-producing cells of the APUD series, and the embryologic, physiologic and pathologic implications of the concept. *J. Histochem. Cytochem.*, **17**, 303–13.

Polak, J. M. and Bloom, S. R. (1979). The diffuse neuroendocrine system. *J. Histochem. Cytochem.*, **27**, 1398–1400.

Polak, J. M. and Bloom, S. R. (1983). Regulatory peptides: key factors in the control of bodily functions. *Br. Med. J.*, **286**, 1461–6.

Polak, J. M. and Marangos, P. J. (1984). In *Evolution and Tumour Pathology of the Neuroendocrine System*. Eds Falkmer, S., Hakanson, R. and Sundler, F., pp. 433–80. Amsterdam, Elsevier Science Publishers.

Polak, J. M. and Varndell, I. M. (Eds) (1984). *Immunolabelling of Electron Microscopy*. Amsterdam, Elsevier Science Publishers.

Polak, J. M. and Bloom, S. R. (Eds) (1985). *Endocrine Tumours: Current Problems in Tumour Pathology*. Edinburgh, Churchill Livingstone.

Polak, J. M. and Van Noorden, S. (Eds) (1986). *Immunocytochemistry: Modern Methods and Applications*, 2nd edn. Bristol, John Wright.

Rozengurt, E. (1986). Early signals in the mitogenic response. *Science*, **234**, 161–6.

Sauville, E. A., Lebagq-Verheyden, A-M., Spindel, E. R., Cuttitta, F., Gazdar, A. D. and Battey, J. F. (1986). Expression of gastrin releasing peptide gene in human small cell lung cancer. *J. Biol. Chem.*, **262**, 2451–7.

Skovgaard Poulson, H., Ozello, L., King, W. J. and Greene, G. L. (1985). The use of monoclonal antibodies to estrogen receptors (ER) for immunoperoxidase detection of ER in paraffin sections of human breast cancer tissue. *J. Histochem. Cytochem.*, **33**, 87–92.

Spindel, E. R., Chin, W. W., Price, J., Rees, L. H., Besser, G. M. and Habener, J. F. (1984). Cloning and characterisation of cDNAs encoding human gastrin-releasing peptide. *Proc. natn. Acad. Sci. USA*, **81**, 5699–703.

Spindel, E. R., Zilberberg, M. D., Habener, J. F. and Chin, W. W. (1986). Two prohormones for gastrin-releasing peptide are encoded by two mRNAs differing by 19 nucleotides. *Proc. natn. Acad. Sci. USA*, **83**, 19–23.

Wilson, B. S. and Lloyd, R. V. (1984). Detection of chromogranin in neuroendocrine cells with a monoclonal antibody. *Am. J. Pathol.*, **115**, 458–68.
Young, W. S. and Kuhar, M. J. (1979). A new method for receptor autoradiography: [^{3}H] opioid receptors in rat brain. *Brain Res.*, **179**, 255–70.

CHAPTER 7

Morphological and biochemical changes in cultured cells exposed to toxic chemicals

P. H. Bach

Introduction

The use of laboratory animals has been an integral part of the screening for toxicity of new chemicals and therapeutic agents (Paget, 1970). Invariably, chemicals damage one (or more) of the major organ systems (the heart, liver, lungs, kidney, etc.) with varying severity. While there has been a tendency in the past just to catalogue these lesions, the rational process of risk assessment (Weil, 1972) now demands that all toxicity should be understood in terms of what might occur in humans exposed to these substances. Thus, there is a need to compare species and to evaluate where extrapolation can be made from animals to man (Parke, 1983). More importantly, we also need to define the molecular basis of chemically-induced lesions (Bridges *et al.*, 1983; Parke, 1983), because a mechanistic understanding of cell injury offers a rational basis for the design of rapid, inexpensive and reliable methods for toxicity screening *in vitro* (Turner, 1983; Williams *et al.*, 1983; Purchase and Conning, 1986). The need to elucidate the molecular mechanisms of these chemically-induced injuries is thus fundamental to understanding all types of organ-selective toxicity, and a variety of 'natural' diseases, including cancer.

Over the last few years it has become apparent that many chemicals target for discrete groups of cells within the kidney, rather than for the whole organ (Bach *et al.*, 1982; Bach and Lock, 1985), although other more extensive degenerative change may subsequently develop (e.g. renal failure—see Bach, 1988). The defining of a molecular mechanism of cellular injury thus depends on the ability to differentiate between the primary pathophysiological events that lead from insult to cell death (or long-term change), as opposed to those peripheral degenerative processes that are not directly involved, or which represent an indirect consequence. The complexity of the intact organ, and especially of the whole animal, has severely limited the identification of most molecular mechanisms of toxicity from studies *in vivo* (Bach and Lock, 1985; Bach, 1988). In order to simplify the experimental system for the study of nephrotoxicity, there has been a steady trend towards the development of *in vitro* methods, where the investigations can be better controlled, and may be performed more rapidly and less expensively. The techniques currently in use, such as renal perfusion, tissue slices, and freshly isolated and cultured cells, have (together with their limitations) recently been reviewed (Bach and Lock, 1982; Bach *et al.*, 1985; Bach and Kwizera, 1988). *In vivo* the numbers of cells that are adversely affected by chemicals are very small compared with the total number of cells in the damaged organ. Thus, in addition to humane reasons for the use of isolated cells and cell cultures, there is a rational need to orientate much of our future research efforts to the study of the small numbers of cells that form the target for chemical injury. Small numbers of cells also form the foci of naturally-occurring diseases, especially cancer. Thus the same principle applies to most other branches of medical research.

Isolated cells have the potential to help identify the primary molecular mechanism(s) of chemical injury and the cascade of degenerative changes that follows (Turner, 1983; Williams *et al.*, 1983; Purchase and Conning, 1986). Many of the methods used to study cells *in vitro* have, however, paralleled the rationale that was originally developed for toxicity testing *in vivo* and make use of either:

1. The *time-average* approach, where a series of discontinuous fixed time point data is used, such as the amount of protein synthesized from a radiolabelled amino acid precursor by different samples of cells, after a fixed period of incubation. This would, for example, be used to compare the rates of protein synthesis in the presence and absence of a chemical that may adversely affect the cells or tissue of interest.
2. A single *end-point* measurement. Here the tissue or cell morphology (or some other variable that is not a time average) is compared with a control sample at a single time point after exposure to a chemical.

These two methods have been the only major 'practical' methods of trying to study chemically-induced lesions in whole animals; and it has generally not been possible to study the dynamics of biochemical changes over a period of time.

Thus far, only relatively few types of selected time-average events (e.g. enzyme leakage from cells, macromolecule or micromolecule synthesis, metabolic degradation, metabolic activation, covalent binding, etc.) and end-point measurements have been studied in isolated and cultured cells, from relatively homogeneous organs such as the liver and skin, or in cultured cell lines, macrophages, sperm, etc. Cells need to be of the same morphological type, otherwise they will not have the same characteristics such as target sensitivity and/or the same profile of biochemical activities, and there will therefore be a fundamental error in the study design. Thus, the more heterogenous the cell populations that are being used, the less likely will it be that the biochemical data generated will be meaningful! The *in vitro* approach of time-average measurement also lacks analytical sensitivity. Typically, 10^4–10^6 cells are needed to provide sufficient material using conventional biochemical techniques, where the response of cells exposed to a test chemical is compared with controls. The use of immunoassays may slightly reduce the number of cells needed. Conventional analytical approaches are especially unsuited to the study of target cell toxicity in heterogeneous organs (such as the kidney and lung), where injury affects one cell type, but not adjacent, morphologically different cells (Bach and Lock, 1985; Boyd, 1980) or in instances where small numbers of isolated cells do not grow rapidly in culture, and are therefore insufficient for the measurement of biochemical changes. Even when cells do grow rapidly, and reach sufficiently large numbers for biochemical evaluation, there may be a substantial de-differentiation, which could change the characteristics of the cells such that they are no longer sensitive to chemicals that damage the original entity *in vivo*.

While it is possible to separate target from non-target cells and to culture these individually, it may be inappropriate to study them in isolation because of the cell–cell interaction in the intact organ. Furthermore, the separation of mixed cells is very complex, and while there are good methods to separate blood cells, this is much more difficult to achieve from solid organs (see Pretlow and Pretlow, 1982–1984). The tissue has to be dispersed into free cells, the identity of which then becomes uncertain (Hay, 1979). Inevitably many cells are damaged by the mechanical or enzymic methods of dispersion (Bach *et al.*, 1985), and the anoxia that develops during some of the separation procedures may compromise the cells' viability. The different approaches used to separate heterogeneous cells are also time consuming, costly, generally of low efficiency and may even yield high proportions of non-viable cells (Ash *et al.*, 1975). In addition, there may still be marked heterogeneity in these 'purified' populations (Scholer and Edelman, 1979) and even in cell lines (Barker and Simmons, 1981; Saier, 1981; Taub and Saier, 1981; Giesen-Crouse *et al.*, 1985). This impure cell population may arise because the criteria used to separate different cells generally make use of only one of a number of determinants. These include size and/or density (Kreisberg *et al.*, 1977), cell surface charge, the presence or absence of surface lectin or antibody-binding sites, attachment and osmotic stability (Bach *et al.*, 1985); all of these are characteristics that may be shared by morphologically different cell types. Mixed populations of cells can be isolated and grown, but, as explained above, conventional biochemical methods cannot be used to study the target cell type in the presence of a non-target population.

Choice of the kidney for toxicity studies *in vitro*

The kidney is made up of over twenty morphologically, biochemically and functionally different cell types. There is also a marked heterogeneity both along individual and between different nephrons (Ross and Guder, 1982). *In vivo* each of the groups of cells form anatomically identifiable regions, such as the glomeruli, the proximal and distal tubules and the medulla; and these may be damaged by selective chemicals *in vivo* (Bach *et al.*, 1982; Bach and Lock, 1985). There are also some chemicals that target very selectively for single renal cell types in each of these discrete regions, and do not affect adjacent and morphologically different cells (Table 7.1). Normally, adjacent but different cell types in the kidney are thought to interact in a variety of ways that are not well understood, but probably include modulation of blood flow, synthesis of substrates and intermediates necessary for the normal function of adjacent cell types, and catabolism of metabolic products of adjacent, but morphologically or biochemically different cell types.

The quest for understanding the biochemical basis of target selective toxicity in the kidney is further complicated by the cascade of degenerative changes that follow the primary injury. Thus, the early target selective (or cell selective) damage to a discrete anatomical region within, say the cortex or medulla, represents the first area of the kidney to be affected. The loss of one part of a cell–cell communication system, the release of autolytic enzymes, the presence of cellular debris in a vessel or tubular lumen, and/or the influence(s) of a variety of other poorly identified factors, result in a secondary series of changes that produce the more widespread organ pathology commonly seen clinically. This includes degeneration leading to acute or chronic renal failure (Bach, 1988). While in many instances a nephrotoxic chemical may affect only a single cell type (as its primary site of action), there are also those compounds that may damage several different cell types, sometimes even in different parts of the kidney (See Table 7.1).

Table 7.1 Chemicals that affect specific regions and/or cell types in the kidney

Chemical	*Target cell type*
Adriamycin	Glomerular epithelial cell
Aminoglycosides	Glomeruli and proximal tubule
p-Aminophenol	Proximal tubule
2-Bromoethanamine	Medullary interstitial cells
Folic acid	Distal tubule
Hexachlorobutadiene (HCBD)	Proximal tubule
HCBD glutathione conjugate	Proximal tubule
HCBD cysteine conjugate	Proximal tubule
HCBD N-acetylcysteine conjugate	Proximal tubule
Mitomycin C	Glomerular mesangial cell
N-Phenylanthranilic acid	Medullary interstitial and proximal tubule cells
Puromycin aminonucleoside	Glomerular epithelial cell
D-Serine	Proximal tubule

At present very little is known about which chemicals reach the kidney, because of the complex interaction of both extra-renal (Parke, 1968; Hucker, 1973) and renal metabolism (Anders, 1980; Rush *et al.*, 1984). More importantly, almost nothing is known about how any one metabolite is distributed at a microscopic level in each of the anatomical regions of the kidney (Mudge, 1985). It has now become recognized that the compartmentalization of the kidney into different regions is very complex due to the paucity of reliable information on how transport processes, trans-cellular pH gradients, solute and/or solvent permselectivity, intracellular binding molecules or organelles and countercurrent trapping or exclusion, may each contribute to the accumulation (or exclusion) of a potential nephrotoxin from any one renal cell type. The available information on the distribution of chemicals within cells is equally unclear, and this is of great significance to the outcome of chemical insult. It is therefore likely that a number of factors contribute individually or collectively to the target selective toxicity, and predetermine where and how given chemicals will produce their nephrotoxic effects (Mudge, 1985). At present, with the exception of the heavy metals (Fowler *et al.*, 1987), very little is known about the sub-cellular distribution of nephrotoxic substances.

Bridging the *in vivo* – *in vitro* gap

There are several factors that make it difficult to distinguish between non-specific cytotoxicity that cannot, and target selective damage that can, be related to damage *in vivo* in the major anatomic regions of damage in the kidney. Any attempt to define the mechanism(s) of target selective toxicity *in vitro* depends on the ability to show, *a priori*, that the same cellular injury occurs specifically in isolated cells. A difficulty associated with studying cultured renal cells is that the disruption of the anatomical integrity, and the cell–cell interactions when renal tissue is dispersed, leads to the disordering of a highly compartmentalized system. The accumulation or exclusion of chemicals from renal cells also makes it difficult to choose concentrations *in vitro* that are realistic, and not inappropriate, to the situation *in vivo*. Under such circumstances, the use of an excessively high concentration of test chemical may cause a cytotoxicity that is not specific to a renal lesion *in vivo*, and the use of an inappropriately low concentration will lead to false negative cytotoxicity data. Finally, both the renal medulla and glomeruli include small numbers of highly differentiated and slow growing cells (Bohman, 1980; Kreisberg and Karnovsky, 1983), and the proximal tubular cells de-differentiate rapidly (Belleman, 1980; Curthoys and Belleman, 1979).

One approach to try to resolve this uncertainty is the parallel use of several different renal cell types to test chemicals that are well known to target for each of these different regions *in vivo*. When these different renal cells types are studied *in vitro* the maintenance of toxic selectivity in isolated cells would imply that the targeting relates to an interaction between the specific cell types and the chemical concerned. Furthermore, the comparison of the concentrations of target selective chemicals needed to produce the adverse effects in cells that are damaged in the whole animal, compared with the concentrations needed to affect other renal cells that are not damaged *in vivo*, offers the potential to compare both the chemicals and the cell types *in vitro* with the effect *in vivo*.

Any attempts to investigate the mechanism of target selective renal injury, with the objective of developing an array of techniques that can be used to screen for nephrotoxicity *in vitro*, must take all of the constraints outlined above into account. We have used freshly isolated glomeruli and proximal tubule rich fragments, as well as cultured cells from these populations and from the medullary interstitial region. Each of the cell types has been exposed to a series of compounds that target either specifically for that region *in vivo* or for the other cells.

We have compared several aspects of renal target cell toxicity to establish the principle that chemical–cell combinations can maintain their target selective toxity *in vitro*:

(i) RENAL MEDULLARY NECROSIS

Renal medullary necrosis follows analgesic or non-steroidal anti-inflammatory drug insult *in vivo* (Bach and Bridges, 1985). Careful time course histopathological studies have shown that it is the renal medullary interstitial cells (RMIC) that undergo necrosis first, while the adjacent collecting duct, covering epithelial and endothelial cells are unaffected. The mechanism(s) for the development of renal papillary necrosis as a result of therapeutic agents has been a focus of speculation for many years, but is still not fully resolved (Bach and Bridges, 1984, 1985). The model papillotoxin, 2-bromoethanamine, is very selective for the medulla, where it causes an acutely developing and dose-related renal papillary necrosis (Bach *et al.*, 1983). The cell type to undergo the earliest degenerative changes are the RMIC (Gregg *et al.*, 1988), as is the case for analgesic and non-steroidal anti-inflammatory drugs; but once again the mechanism(s) are poorly understood.

We have cultured rodent RMIC and shown them to be sensitive to established papillotoxins such as 2-bromoethanamine and 3-bromopropanamine, but not to the non-papillotoxic fluoroethanamine (Benns *et al.*, 1985). Similarly, paracetamol (acetaminophen) is much more cytotoxic to these cells than are its metabolites (Benns *et al.*, 1985), but the cultured RMIC are not sensitive to the chemicals that damage the glomeruli or the proximal tubule (see over).

(ii) GLOMERULI

In vivo, glomerular epithelial (but not mesangial) cells are injured selectively by both adriamycin (doxorubicin) (Bertani *et al.*, 1982) and puromycin aminonucleoside (PAN) (Brenner *et al.*, 1977), a lesion that causes a model nephrotic syndrome.

Glomeruli are the most easily isolated renal cell type, and accordingly, they have been studied, characterized and cultured (Holdsworth *et al.*, 1978; Kreisberg *et al.*, 1978; Scheinman and Fish, 1978; Foidart *et al.*, 1979, 1980, 1981; Morita *et al.*, 1980; Oite *et al.*, 1980; Striker *et al.*, 1980; Kreisberg and Karnovsky, 1983). Freshly isolated glomeruli are sensitive to therapeutically used chemicals (Meezan and Brendel, 1973; Savin *et al.*, 1985) and show biochemical changes when isolated from animals with a model nephrotic syndrome (Chow and Drummond, 1969). Studies *in vitro* have shown that PAN produces its primary effect by damaging the epithelial cells from cultured glomeruli (Kreisberg and Karnovsky, 1983).

We have used freshly isolated glomeruli and shown that whereas adriamycin (doxorubicin) (<0.05 mM) inhibits amino acid incorporation into glomerular basement membrane (Ahmed, I., unpublished results), proximal tubular fragments are only affected at 0.5 mM, and adriamycin has no effect on the RMIC (Ketley, C. P., unpublished results). The mechanism for PAN-induced glomerular epithelial cell necrosis is not understood, but may, in common with adriamycin, be related to redox cycling (Gianni *et al.*, 1983) or to prostaglandin synthesis and/or related peroxidative metabolism (Schlondorff *et al.*, 1980; Sraer *et al.*, 1980, 1983; Stollenwerk Petrulis *et al.*, 1981). More importantly, the glomeruli are unaffected by the haloalkene metabolite that damages proximal tubular cells (see below) and to the papillotoxin 2-bromoethanamine (Bach *et al.*, 1986; Ketley, C. P., Ahmed, I., Dixit, M. and Dela Cruz, L., unpublished results).

(iii) PROXIMAL TUBULES

The proximal tubule is selectively damaged *in vivo* by a number of different classes of chemicals, one of the most target selective of which is hexachlorobutadiene (HCBD). This haloalkene is metabolized in the liver to the glutathione conjugate, which is converted in the kidney to the N-acetylcysteine conjugate, the reactive metabolite (Lock, 1982; Nash *et al.*, 1984).

Renal cortical tubules have been isolated (Ormstad *et al.*, 1981) and characterized in terms of biochemistry (Dawson, 1972, 1975; Moldeus *et al.*, 1978, 1980; Ormstad, 1987), organic ion transport (Rylander-Yueh *et al.*, 1985) and drug metabolism (Fry *et al.*, 1978; Moldeus *et al.*, 1978; Fry and Perry, 1981; Ormstad, 1988). Isolated tubules have been used to show the adverse effects *in vitro* of HCBD metabolites (Earl, 1985), branched chain alpha-ketoacids (Stumpf and Kraus, 1978) and antibiotics (Viano *et al.*, 1983; Savin *et al.*, 1985). Rat cortical cells have also been cultured, but while these may have had proximal tubular characteristics, they were undoubtedly mixed populations (Curthoys and Belleman, 1979; Belleman, 1980).

We have shown the active metabolite HCBD–N-acetylcysteine (HCBD–NaC) to inhibit the incorporation of amino acids into proximal tubular macromolecules *in vitro* (Ketley and Bach, 1988a, b). Probenecid has already been shown to protect animals exposed to HCBD or HCBD–NaC *in vivo* (Lock, 1985). In our studies, pre-treatment of the tubular fragment system with 400 μM probenecid before exposure to HCBD–NaC restored 95% of their macromolecular synthetic capacity (Ketley and Bach, 1988b). HCBD–NaC targets vary selectively for cultured proximal tubular cells (Ketley, C. P., unpublished results), but has no adverse effect on fresh or cultured glomeruli (Ahmed, I., unpublished results). Several other classical proximal tubular toxins also produce similar effects *in vitro* (Ketley, C. P., unpublished results; Kwizera and Bach, 1988), and a significant degree of *in vitro* selectivity is supported by the fact that the proximal tubular cells are only adversely affected by very high concentrations of the model papillotoxin 2-bromoethanamine and by adriamycin.

In summary, the model lesions caused by 2-bromoethanamine, adriamycin and the acetylcysteine metabolite of hexachlorobutadiene are particularly relevant *in vivo*, because primary lesions occur in single cell types, which then give rise to a series of degenerative effects and may also cause extra-renal changes. Thus, *in vivo* HCBD causes an acute proximal tubular lesion which repairs very rapidly, adriamycin damages the glomerular epithelial cells, leading to a slow progressive nephrotic syndrome that may eventually cause changes in the liver, and bromoethanamine-induced renal medullary necrosis is a focal damage to the renal medullary interstitial cells that is followed by a cascade of events that include marked cortical degeneration, renal failure, upper urothelial hyperplasia and carcinoma, together with a large number of extra-renal abnormalities. Having bridged the *in vivo* – *in vitro* gap for isolated RMIC, glomerular and proximal tubular cells (whereas chemicals which do not damage these cells *in vivo* also have no effects *in vitro*), we then wished to define the mechanisms involved.

Micromethods for studying the biochemistry of single cells

A variety of micromethods (Neuhoff, 1973) such as microdissection (using nanogram quantities of tissue) has been used to study specific enzyme activities by 'cycling' radiolabelled or fluorescent assays. By contrast, cytochemistry typically uses 10–10^3 cells, which suffice for quantitation (Chayen, 1985).

High-resolution microscopy assessment of live cells

High-resolution inverted microscopy can be used to circumvent many of the problems that currently limit those aspects of the use of isolated cells to study subcellular mechanisms of 'cytotoxicity'. This microscopic approach makes use of several optical systems, in addition to conventional and phase-contrast illumination. Normarski (differential interference contrast) optics offer excellent image contrast and detail, in unstained live cells, and at a resolution greater than $0.2\,\mu$m. A single plane of the cell is viewed, thus interference from other cellular material is minimal, and nuclei and their contents, intra-cellular organelles, aspects of the cytoskeleton and cell surface features of the living cells can be studied (Beneke, 1966). Without Normarski optics, similar data can only be acquired from the differential staining of fixed (dead) cells, or by the use of ultrastructural methods. The highest magnifications using high-resolution microscopy are comparable with low-power scanning and transmission electron microscopy, but can be applied to live cells. This is a most important and time-saving technique of establishing where chemically-induced cell changes warrant further ultrastructural evaluation.

High-resolution microscopy can also be both qualitative and quantitative, by using

Table 7.2 Advantages of videomicrography for cell monitoring

Video cameras

Resolution: Greater than 0.2 μm, which is the resolution of UV light micrography

Sensitivity: Choice of camera video tube and enhancing system provides sensitivity down to single photons. Spectral response 140 to 850 nm; can be controlled by filters.

Image manipulation

Electronic manipulation allows for contrast enhancement, vertical and horizontal shading correction, white clipping, inversion, boosting, rolling averages, photon counting, mottle subtraction, pseudo-colour, colour enhancement

Computer control of system

Measurements of distance between two points, areas, size distribution, tracking movement, intensity display, morphometric analysis

Time-lapse monitoring

highly-selective fluorescent probes which can be quantitated at $< 10^{-12}$ mol (Udenfriend, 1969). Finally, the use of time-lapse video recording (to study the sequence of sub-cellular changes) can be used to study time-course changes and recovery of both surface features and intracellular organelles within live cells. More importantly, this technique provides *dynamic data* on the response of *live* cells (Parsons, 1986) by video recording of cellular responses to chemicals.

Video monitoring and time-lapse recording

VIDEO MONITORING

Over the last few years, the increased use of video monitoring of cell cultures has added several new and important dimensions to the type of information that can be gathered from studying cell changes (Allen and Allen, 1982) and especially and most recently the cellular response to a chemical insult (Bach, P. H., unpublished results; Parsons, 1986). The advantages of video monitoring over the human eye and/or conventional photographic emulsion imaging are listed in Table 7.2. Unquestionably, the most important advantages of the video system are the several orders of magnitude increase in sensitivity to light and the ability to manipulate the images in 'real time'. Video imaging has, in fact, now superseded the optical resolution of the light microscope, and a new generation of instrument design is needed to take full advantage of the full potential of electronic visualization.

TIME-LAPSE MONITORING OF CELL CULTURES

One of the major disadvantages of not using time-lapse monitoring of cells in culture is the inability to comprehend a focal loss of cells while the remaining population grows over into the 'spaces' that have developed (a phenomenon that is, incidently, comparable with superinfection overgrowth by bacteria following broad-spectrum antibiotic usage). Thus, a slow rate of cell death in a heterogeneous population that contains resistant as well as sensitive species could go undetected. The marked uncertainty with regard to the sequence of events that follow cell exposure to a chemical is also a disadvantage because morphological changes could give a very clear indication of any molecular changes that may be occurring. Thus, for example, rounding and detachment of cells (or altered mobility), together with membrane blebbing may indicate energy-related or cytoskeletal changes, whereas altered lysosomal, mitochondrial or nuclear integrity imply direct injury to these organelles.

The dichotomy of structure and biochemical function can be bridged by the use of cytochemical methods to monitor changes in a small number of cells exposed to toxins. Cytochemistry can be used to study the distribution of selected enzymes and enzyme systems, and of other biochemical processes in individual cultured cells (Chayen, 1985). The energy-dependent uptake of phenol red into the proximal tubule was one of the very early scientific methods used to assess the active transport functions of this part of the nephron (Forster and Taggart, 1950). The use of visible chromophores is, however, limited to a few vital stains, and thus the number of metabolic processes that can be assessed are therefore rather restricted.

Fluorescent probes

Fluorescent probes are much more versatile and can be used to study a diverse array of biochemical processes. Also these fluorophores offer several orders of magnitude more sensitivity with which to link morphological change to biochemical events. Discrete biochemical perturbations in specific organelles, membranes or the cytoplasm (see Table 7.3) can be linked to sub-cellular morphological changes. The origin of any one cell type in a mixed population can be established by using antibody-tagged probes or other selective biochemical criteria such as specific enzymes, large numbers of mitochondria, etc. Changes in adjacent but different cell types can be compared and contrasted; and this is especially important where one cell type forms the target for chemical injury *in vitro* and the other(s) do not.

The use of fluorescent probes and time-lapse video to link morphological degeneration and biochemical mechanisms of renal cell injury

RENAL MEDULLARY INTERSTITIAL CELLS

Fluorescent probes have revealed high levels of peroxidase activity and/or peroxides in RMIC (Homan-Muller, 1975), and the use of Nile Red (Greenspan *et al.*, 1985) can confirm the presence of lipid droplets (Bohman, 1980) in these cells, which are known to contain very high levels of polyunsaturated fatty acids (Bojsen, 1980). The mechanism of analgesic-induced renal medullary necrosis is thought to be linked to the peroxidative activation of agents to form reactive intermediates (Zenser *et al.*, 1979; Mohandas *et al.*, 1981; Bach and Bridges, 1984, 1985). In the presence of high levels of polyunsaturated fatty acids, the reactive analgesic intermediate could cause lipid peroxidation and lead to cell

Table 7.3 Selective fluorescent probes that allow subcellular biochemistry and morphology to be interlinked

System being assessed	*Flourescent probes*
Enzyme probes	
Esterases	Fluorescein diacetate
Mixed function oxidase	Ethoxyresorufin
Peroxidase activities	2′,7′-Dichloro-dihydrofluorescein
Transport probes	
Cationic transport system	4-Acetamido-4′-isothiocyanato-stilbene-2,2′-disulphonic acid
Organelle probes	
Mitochondria	Dimethyl-aminostyryl-methylpyridinium-iodide
Lysosomes	FluoroBora I
Golgi apparatus	FluoroBora II
Nucleus	FluoroBora T
Cytoplasmic probes	
Hydrophobic and hydrophilic regions	FluoroBora P
Ca^{2+} influx	Quin 2
Lipid probes	
Cellular and organelle membranes	5-(N-hexadecanoyl)aminoeosin
Neutral lipid droplets	Nile Red
Immunological probes	
Cytoskeletal elements	Polyclonal and monoclonal antibodies tagged with a fluorophore.
Selective binding proteins	
Microtubular system	Phalloidin tagged with a fluorophore.
Carbohydrate constituents	Lectins tagged with a fluorophore.
Drugs	
	Tetracycline
	Chloroquine
	Adriamycin
Natural products	
	Vitamin A
	Conjugated dienes
	Lipofuscin

death (Bach and Bridges, 1984). Recent time-lapse studies on RMIC indicate that those cells with most lipid droplets are the most sensitive to 2-bromoethanamine (Bach, P. H., unpublished results). In an attempt to confirm an association between peroxidase activity and lipid material, we have also studied the sensitivity of cell lines to papillotoxins. Both RMIC and 3T3 fibroblasts were sensitive to 2-bromoethanamine (Bach *et al.*, 1986), but there were no cytotoxic changes in MDCK or HaK cells exposed to a 10-fold increased concentration of 2-bromoethanamine for four times as long. RMIC are thought to be fibroblastic in origin (Bohman, 1980), hence the comparison with 3T3 cells, whereas HaK (Hull *et al.*, 1976) and MDCK (Herzlinger *et al.*, 1982; Hassid, 1983) represent proximal and distal renal epithelial cells, respectively. The absence of lipid droplets and peroxidase activity from HaK and MDCK cells may explain the lack of 2-bromoethanamine cytotoxicity.

ADRIAMYCIN-INDUCED GLOMERULAR INJURY

Fluorescent illumination of cultured glomeruli (with outgrowths of both mesangial and epithelial cells) indicate the preferential uptake of adriamycin into epithelial cell nuclei, but not mesangial cell nuclei. Time-lapse recording shows that nuclear degeneration precedes cell death in these cells (Bach *et al.* 1986).

The potential of the single-cell approach to alternatives in toxicology

The future uses of the single-cell approach to understand the mechanisms of selective injury and general cytotoxicity (and hence develop rational *in vitro* tests for evaluating chemical safety and improving drug design) are exciting. Most of the 'single-cell' techniques that have been described in this chapter have been a part of the cell biologist's armamentarium for over twenty years. Recent developments in cell biology may be especially appropriate for applications *in vitro*; these include the following:

(i) The application of patch clamping techniques. These will allow us to gain a greater insight into the electrophysiology of cells and membrane changes associated with chemical insult (Hamill *et al.*, 1981). The patch clamp technique is especially useful for studying ionic channels and membrane currents. This can be undertaken while fluorescent probes are used concomitantly to study the biochemical changes taking place in the same cells.

(ii) Computer-based image analysis of the movement and changes in shapes of cells will allow us to understand better the role played by the cytoskeleton in normal cell function, and how this is altered when cells are exposed to target-specific chemicals.

(iii) Cell fusion is a powerful technique that allows novel characteristics to be combined. Frog renal epithelial cells (diameter $\pm 10\,\mu$m) have been fused (using polyethylene glycol) with 'giant cells' (diameter $\pm 100\,\mu$m) from the diluting segment to facilitate direct electrode studies (Oberleithner *et al.*, 1986). Similarly, the electrofusion of cells that will permit the combination of features unique to two different cells (Zimmerman *et al.*, 1981; Zimmerman, 1982) that are not normally present *in vitro*. Thus cell lines could be combined by one of these standard cell-fusion techniques to give, for example, a transport process of one cell type (for example, that in HaK or MDCK cells) and a metabolic activation system (such as a peroxidase or P-450 system in fibroblasts or hepatocytes, respectively) to test the hypothesis that both characteristics must be present in a particular cell type for a specific target toxicity to occur.

The use of human tissue

There are several reasons to try to make use of human renal tissue as the source of cells for toxicity studies *in vitro*. The rat kidney differs very significantly from that of humans (Mudge, 1982). More importantly, most experimental animals are free of pre-existing renal disease, but this is not the case in man, where decreased functional reserve may be one of the least adequately studied aspects of comparative nephrotoxicity. Thus the use of human renal cells offers the potential to circumvent the need to extrapolate animal studies *in vitro* to man. There have already been a number of published studies on human kidneys. These have been characterized in terms of normal and diseased tissue (Roth *et al.*, 1979; Pellett *et al.*, 1984) and glomeruli (Holdsworth *et al.*, 1978; Scheinman and Fish, 1978; Striker *et al.*, 1980; Ardaillou *et al.*, 1983; Sraer *et al.*, 1983). Recently, there has been a trend towards studying cultured human renal proximal tubular cells exposed to nephrotoxins (Chatterjee *et al.*, 1985; Trifillis *et al.*, 1985a, b).

Our own research programme is following a similar progression. We have established an experimental basis to study the molecular mechanisms of target selective renal injury, and confirmed that the *in vivo–in vitro* gap can be bridged for discrete cell types in the rat kidney. It is now possible to extend these investigations into the study of other species. We have started on the pig kidney, which is morphologically and functionally very similar to the human kidney (Mudge, 1982). Furthermore, large quantities of disease-free tissue are readily available from abattoir material, thus obviating the need to sacrifice laboratory animals. We have already shown that pig glomeruli are very sensitive to adriamycin (Dela Cruz, L., unpublished results), but not to those chemicals that damage other parts of the kidney. In the long-run it should also be possible to apply similar studies to renal cells from human tissue.

Conclusion

The use of the single-cell approach offers a very powerful potential to study intracellular events and to assess sub-cellular morphology and functions. Once these are established, there is a rational basis to screen for similar types of toxicity, either using isolated cells from animals, or human tissue to circumvent the uncertainties associated with extrapolating from animals. If there is a clear-cut molecular sequence of events, such as the peroxidative activation of papillotoxic drugs and lipid peroxidation (assuming that this were the mechanism), then the use of 3T3 cell lines might be appropriate. Alternatively, a totally artificial system containing purified horseradish peroxidase activates the several analgesics that cause renal papillary necrosis *in vivo* to form reactive intermediates that bind covalently to macromolecules (Zenser *et al.*, 1979; Mohandas *et al.*, 1981). This simple, rapid and inexpensive cell-free system may finally provide the ultimate alternative strategy.

Summary

Many chemicals target selectively for one of the major organ systems (such as the kidney) and damage discrete cell types, whereas adjacent, but morphologically different cells, are not affected. The elucidation of the mechanism(s) of these unique interactions between the 'target cell' and the toxin has proved most difficult *in vivo*. Cell culture offers an ideal way to study the molecular basis of selective cell injury, provided the targeting is maintained *in vitro*.

A number of chemicals do maintain their targeting towards isolated and cultured renal glomerular epithelial cells, proximal tubular cells and renal medullary interstitial cells. Most *in vitro* methods have, however, been severely restricted because they require 10^4–10^6 cells if changes are to be assessed by conventional biochemical methods. Microscopy with special illumination, fluorescent probes, and time-lapse recording circumvent many of the existing constraints. Less than a hundred cells can be used to follow the sequence of subcellular morphological aberration and the dynamics of biochemical changes in *live* cells exposed to target-selective toxins. This approach can be applied to those systems where the *in vivo–in vitro* gap has been bridged, and specifically serves the purpose of linking the dichotomy between cell morphology and cell biochemistry, and allows mechanistic changes to be followed in the presence of target selective chemicals.

Acknowledgements

The research was supported by the Humane Research Trust, the John Hopkins Center for Alternatives to Animals in Testing, the Dr Hadwen Trust for Humane Research and, in part, by the Wellcome Trust and the Commission of the European Communities. I am grateful to Dr E. A. Lock for the generous gifts of hexachlorobutadiene metabolites, M. E. van Ek for preparing the manuscript and to Dr E. N. Kwizera for his critical comments.

References

Allen, R. D. and Allen, N. S. (1982). Video-enhanced microscopy with a computer frame memory. *J. Microscopy*, **129**, 3–17.

Anders, M. (1980). Metabolism of drugs by the kidney. *Kidney int.*, **18**, 636–47.

Ardaillou, N., Nivez, M.-P., Striker, G. and Ardaillou, R. (1983). Prostaglandin synthesis by human glomerular cells in culture. *Prostaglandins*, **26**, 773–84.

Ash, S. R., Cuppage, F. E., Hodes, M. E. and Selkurt, E. E. (1975). Culture of isolated renal tubules: A method of assessing viability of normal and damaged cells. *Kidney int.*, **7**, 55–60.

Bach, P. H. (1988). Towards the sensitive and selective diagnosis of chemically induced renal injury. *Xenobiotica*, **18**, 685–98.

Bach, P. H. and Lock, E. A. (1982). The use of renal tissue slices, perfusion and infusion techniques to assess renal function and malfunction. In *Nephrotoxicity: Assessment and Pathogenesis*, Eds Bach, P. H., Bonner, F. W., Bridges, J. W. and Lock, E. A., pp. 128–43. Chichester, J. Wiley.

Bach, P. H. and Bridges, J. W. (1984). The role of prostaglandin synthetase mediated metabolic activation of analgesics and non-steroidal anti-inflammatory drugs in the development of renal papillary necrosis and upper urothelial carcinoma. *Prostaglandins and Leukotrienes in Medicine*, **15**, 251–74.

Bach, P. H. and Bridges, J. W. (1985). Chemically induced renal papillary necrosis and upper urothelial carcinoma. *CRC crit. Rev. Toxicol.*, **15**, 217–439.

Bach, P. H. and Lock, E. A. (Eds) (1985). *Renal Heterogeneity and Target Cell Toxicity*. Chichester, J. Wiley.

Bach, P. H. and Kwizera, E. N. (1988). Nephrotoxity: A rational approach to *in vitro* target cell injury in the kidney. *Xenobiotica*, in press.

Bach, P. H., Bonner, F. W., Bridges, J. W. and Lock, E. A. (Eds) (1982). *Nephrotoxicity: Assessment and Pathogenesis*. Chichester, J. Wiley.

Bach, P. H., Grasso, P., Molland, E. A. and Bridges, J. W. (1983). Changes in the medullary glycosaminoglycan histochemistry and microvascular filling during the development of 2-bromoethanamine hydrobromide-induced renal papillary necrosis. *Toxicol. appl. Pharmacol.*, **69**, 333–44.

Bach, P. H., Ketley, C. P., Benns, S. E., Ahmed, I. and Dixit, M. (1985). The use of isolated and cultured renal cells in nephrotoxicity—Practice, potential and problems. In *Renal Heterogeneity and Target Cell Toxicity*, Eds Bach, P. H. and Lock, E. A., pp. 505–18. Chichester, J. Wiley.

Bach, P. H., Ketley, C. P., Dixit, M. and Ahmed, I. (1986). The mechanisms of target cell injury in nephrotoxicity. *Food Chemical Toxicol.*, **24**, 775–9.

Barker, G. and Simmons, N. L. (1981). Identification of two strains of cultured canine renal epithelial cells (MDCK cells) which display entirely different physiological properties. *Quart. J. exp. Physiol.*, **66**, 61–72.

Belleman, P. (1980). Primary monolayer cultures of liver parenchymal cells and kidney tubules as a useful new model for biochemical pharmacology and experimental toxicology. Studies *in vitro* on hepatic membrane transport, induction of liver enzymes and adaptive changes in renal cortical enzymes. *Arch. Toxicol.*, **44**, 63–84.

Beneke, G. (1966). Application of interference microscopy to biological material. In *Introduction to Quantitative Cytochemistry*, Ed. Wied, G. L., pp. 63–92. New York, Academic Press.

Benns, S. E., Dixit, M., Ahmed, I., Ketley, C. P. and Bach, P. H. (1985). The use of renal medullary cell cultures as alternatives to live animals for studying renal medullary toxicity. In *Alternative Methods in Toxicology*, Ed. Goldberg, A., Vol. 3, pp. 435–47. Baltimore, M. A. Liebert.

Bertani, T., Poggi, A., Pozzoni, R., Delaini, F., Sacchi, G., Thoua, Y., Mecca, G., Remuzzi, G. and Donati, M. B. (1982). Adriamycin-induced nephrotic syndrome in rats. *Lab. Invest.*, **46**, 16–23.

Bohman, S.-O. (1980). The ultrastructure of the renal medulla and the interstitial cells. In *The Renal Papilla and Hypertension*, Eds Mandal, A. K. and Bohman, S.-O., pp. 7–34. New York, Plenum Publishing Corp.

Bojsen, I. N. (1980). Fatty acid composition and depot function of lipid droplet triacylglycerols in renomedullary interstitial cells. In *The Renal Papilla and Hypertension*, Eds Mandal, A. K. and Bohman, S.-O., pp. 121–47. New York, Plenum Publishing Corp.

Boyd, M. G. (1980). Biochemical mechanisms in chemically induced lung injury: Roles of metabolic activation. *CRC crit. Rev. Toxicol.*, **7**, 103–76.

Brenner, B. M., Bohrer, M. P., Baylis, C. and Deen, W. M. (1977). Determinants of glomerular permselectivity. Insights derived from observations *in vivo*. *Kidney Int.*, **12**, 229–37.

Bridges, J. W., Benford, D. J. and Hubbard, S. A. (1983). Mechanisms of toxic injury. *Ann. N.Y. Acad. Sci.*, **407**, 42–63.

Chatterjee, S., Trifillis, A. L. and Regec, A. L. (1985). Morphological and biochemical effects of gentamicin on cultured human renal tubular cells. In *Renal heterogeneity and Target Cell Toxicity*, Eds Bach, P. H. and Lock, E. A., pp. 549–52. Chichester, J. Wiley.

Chayen, J. (1985). Quantitative cytochemistry: A precise form of cellular biochemistry. *Biochem. Soc. Trans.*, **12**, 887–98.

Chow, A. Y. K. and Drummond, K. N. (1969). Incorporation and hydroxylation of [^{3}H]-3-4-proline as an index of glomerular basement membrane synthesis in normal and nephrotoxic nephritic rats. *Lab. Invest.*, **20**, 213–18.

Curthoys, N. P. and Bellemann, P. (1979). Renal cortical cells in primary monolayer culture. Enzymatic changes and morphological observations. *Expl Cell Res.*, **121**, 31–45.

Dawson, A. G. (1972). Preparation and some properties of a suspension of fragmented tubules from the rat kidney. *Biochem. J.*, **130**, 525–32.

Dawson, A. G. (1975). Effects of acetylsalicylate on gluconeogenesis in isolated rat kidney tubules. *Biochem. Pharmacol.*, **24**, 1407–11.

Earl, L. K. (1985). Experimental Studies on Mechanisms of Nephrotoxicity. PhD thesis, University of London.

Foidart, J. B., Dechenne, C. A., Mahieu, P., Creutz, C. E. and De Mey, J. (1979). Tissue culture of normal rat glomeruli: Isolation and morphological characterization of two homogeneous cell lines. *Invest. Cell Pathol.*, **2**, 15–26.

Foidart, J. B., Dubois, C. H. and Foidart, J.-M. (1980). Tissue culture of normal rat glomeruli: Basement membrane biosynthesis by homogeneous epithelial and mesangial cell lines. *Int. J. Biochem.*, **12**, 197–202.

Foidart, J. B., Dechenne, C. A. and Mahieu, P. (1981). Tissue culture of normal rat glomeruli: Characterization of collagenous and non-collagenous basement membrane antigens on the epithelial and mesangial cells. *Diagnostic Histopath.*, **4**, 71–7.

Forster, R. P. and Taggart, J. V. (1950). Use of isolated renal tubules for the examination of metabolic processes associated with active cellular transport. *Cell. Compar. Physiol.*, **36**, 251–70.

Fowler, B. A., Mistry, P. and Goering, P. L. (1987) Mechanisms of metal-induced nephrotoxicity. In *Nephrotoxicity in the Experimental and the Clinical Situation*, Eds Bach, P. H. and Lock, E. A., pp. 659–83. Dordrecht, Nijhoff.

Fry, J. R. and Perry, N. K. (1981). The effect of Aroclor-1254 pretreatment on the phase I and phase II metabolism of 7-ethoxycoumarin in isolated viable rat kidney cells. *Biochem. Pharmacol.*, **30**, 1197–201.

Fry, J. R., Wiebkin, P., Kao, J., Jones, C. A., Gwynn, J. and Bridges, J. W. (1978). A comparison of drug-metabolising capability in isolated viable rat hepatocytes and renal tubule fragments. *Xenobiotica*, **8**, 113–20.

Gianni, L., Corden, B. J. and Myers, C. E. (1983). The biochemical basis of anthracycline toxicity and anti-tumor activity. *Rev. Biochem. Toxicol.*, **5**, 1–82.

Giesen-Crouse, E. M., Imbs, J. L., Steffan, A., Schmidt, M. and Schwartz, J. (1985). Characterization of MDCK subcultures with respect to adenylate cyclase induction. In *Renal Heterogeneity and Target Cell Toxicity*, Eds Bach, P. H. and Lock, E. A., pp. 519–22. Chichester, J. Wiley.

Greenspan, P., Mayer, E. P. and Fowler, S. D. (1985). Nile Red: A selective stain for intracellular lipid droplets. *J. Cell Biol.*, **100**, 965–73.

Gregg, N., Courtauld, E. A., and Bach, P. H. (1988). High resolution light microscopic morphological and microvascular changes in an acutely-induced renal papillary necrosis. *Pathol Toxicol.*, submitted.

Hamill, O. P., Marty, A., Neher, E., Sakmann, B. and Sigworth, F. J. (1981). Improved patch-clamp techniques for high resolution current recording from cell and cell-free membrane patches. *Pflügers Arch.*, **391**, 85–100.

Hassid, A. (1983). Modulation of cyclic 3′5′-adenosine monophosphate in cultured renal (MDCK) cell by endogenous prostaglandins. *J. Cell. Physiol.*, **116**, 297–307.

Hay, R. J. (1979). Cells for culture. In *Cell Populations*, Ed. Reid, E., pp. 143–60. Chichester, Ellis Horwood.

Herzlinger, D. A., Easton, T. G. and Ojakian, G. K. (1982). The MDCK epithelial cell line expresses a cell surface antigen of the kidney distal tubule. *J. Cell Biol.*, **93**, 269–77.

Holdsworth, S. R., Glasgow, E. F., Atkins, R. C. and Thomson, N. M. (1978). Cell characteristics of cultured glomeruli from different animal species. *Nephron*, **22**, 454–9.

Homan-Muller, J. W. T., Weening, R. S. and Roos, D. (1975). Production of hydrogen peroxide by phagocytozing human granulocytes. *J. Lab. clin. Med.*, **85**, 198–207.

Hucker, H. B. (1973). Intermediates in drug metabolism reactions. *Drug Metab. Rev.*, **2**, 33–54.

Hull, R. N., Cherry, W. R. and Weaver, G. W. (1976). The origin and characteristics of a pig kidney cell strain, LLC-PK_1. *In vitro*, **12**, 670–7.

Ketley, C. P. and Bach, P. H. (1988a). The effects of target selective nephrotoxins on the incorporation of proline into rat renal tubular fragment macromolecules. Submitted.

Ketley, C. P. and Bach, P. H. (1988b). The effects of haloalkene cysteine analogues on the incorporation of amino acids into rodent proximal tubular fragment macromolecules. Submitted.

Kreisberg, J. I. and Karnovsky, M. J. (1983). Glomerular cells in culture. *Kidney int.*, **23**, 439–47.

Kreisberg, J. I., Pitts, A. M. and Pretlow, T. G. (1977). Separation of proximal tubule cells from suspensions of rat kidney cells in density gradients of Ficoll in tissue culture medium. *Am. J. Pathol.*, **86**, 591–600.

Kreisberg, J. I., Hoover, R. L. and Karnovsky, M. J. (1978). Isolation and characterization of rat glomerular epithelial cells *in vitro*, *Kidney int.*, **14**, 21–30.

Kwizera, E. N. and Bach, P. H. (1988). The effects of aminoglycosides on the incorporation of amino acids into macromolecules of isolated rat renal proximal tubules. *ATLA*. Submitted.

Lock, E. A. (1982). Renal necrosis produced by halogenated chemicals. In *Nephrotoxicity: Assessment and Pathogenesis*, Eds Bach, P. H., Bonner, F. W., Bridges, J. W. and Lock, E. A., pp. 396–408. Chichester, J. Wiley.

Lock, E. A. (1985). Probenecid prevents the nephrotoxicity provoked by the N-acetylcysteine conjugate of hexacloro-1: 3-butadiene. In *Renal Heterogeneity and Target Cell Toxicity*, Eds Bach, P. H. and Lock, E. A., pp. 149–52. Chichester, J. Wiley.

Lock, E. A., Odum, J. and Ormond, P. (1986). Transport of N-acetyl-S-pentachloro-l, 3-butadienylcysteine by the renal cortex. *Arch. Toxicol.*, **59**, 12–15.

Meezan, E. and Brendel, K. (1973). Effect of ethacrynic acid on oxidative metabolism in isolated glomeruli. *J. Pharmac. exp. Ther.*, **187**, 352–64.

Mohandas, J., Duggin, G. G., Horvath, J. S. and Tiller, D. J. (1981). Metabolic oxidation of acetaminophen (paracetamol) mediated by cytochrome P-450 mixed function oxidase and prostaglandin endoperoxide synthase in rabbit kidney. *Toxicol. appl. Pharmacol.*, **61**, 252–9.

Moldeus, P., Jones, D. P., Ormstad, K. and Orrenius, S. (1978). Formation and metabolism of a glutathione-S-conjugate in isolated rat liver and kidney cells. *Biochem. Biophys. Res. Commun.*, **83**, 195–200.

Moldeus, P., Ormstad, M. and Reed, D. J. (1980). Turnover of cellular glutathione in isolated rat-kidney cells. *Eur. J. Biochem.*, **116**, 13–16.

Morita, T., Oite, T., Kihara, I., Yamamoto, T., Hara, M., Naka, A. and Ohno, S. (1980). Culture of isolated glomeruli from normal and nephritic rabbits. I. Characterization of outgrowing cells. *Acta pathol. jpn.*, **30**, 917–26.

Mudge, G. H. (1982). Comparative pharmacology of the kidney: Implications for drug-induced renal failure. In *Nephrotoxicity: Assessment and Pathogenesis*, Eds Bach, P. H., Bonner, F. W., Bridges, J. W. and Lock, E. A., pp. 504–18. Chichester, J. Wiley.

Mudge, G. H. (1985). Pathogenesis of nephrotoxicity: pharmacological principles. In *Renal Heterogeneity and Target Cell Toxicity*, Eds. Bach, P. H. and Lock, E. A., pp. 1–12, Chichester, J. Wiley.

Nash, J. A., King, L. J., Lock, E. A. and Green, T. (1984). The metabolism and disposition of hexachloro-l,3-butadiene in the rat and its relevance to nephrotoxicity. *Toxicol. appl. Pharmacol.*, **73**, 124–37.

Neuhoff, V. (Ed.). (1973). *Micromethods in Molecular Biology*, London, Chapman & Hall.

Oberleithner, J., Schmidt, B. and Dietl, P. (1986). Fusion of renal epithelial cells: A model for studying cellular mechanisms of ion transport *Proc. natn. Acad. Sci. USA*, **83**, 3547–51.

Oite, T., Morita, T., Kihara, I., Yamamoto, T., Suzuki, Y. and Okada, K. (1980). Culture of isolated glomeruli from normal and nephritic rabbits. II. Kinetics of outgrowing cells. *Acta pathol. jap.*, **30**, 927–35.

Ormstad, K. (1987). Metabolism of glutathione in the kidney. In *Nephrotoxicity in the experimental and the Clinical Situation*, Eds Bach, P. H. and Lock, E. A., pp. 405–28. Dordrecht, Nijhoff.

Ormstad, K., Jones, D. P. and Orrenius, S. (1981). Preparation and characterisation of isolated kidney cells. *Meth. Enzym.*, **77**, 137–46.

Paget, G. E. (1970). *Methods in Toxicology*. Oxford, Blackwells.

Parke, D. V. (1968). *Biochemistry of Foreign Compounds*. Oxford, Pergamon Press.

Parke, D. V. (1983). A more scientific approach to the safety evaluation of chemicals. In *Animals in Scientific Research: An Effective Substitute for Man?*, Ed. Turner, P., pp. 7–28. London, Macmillan Press.

Parsons, J. F. (1986). Toxic Effects in Macrophages and Mast Cell Preparations Perfused *in vitro*. PhD thesis, University of Bradford.

Pellett, O. L., Smith, M. L., Thoene, J. G., Schneider, J. A. and Jonas, A. J. (1984). Renal cell culture using autopsy material from children with cystinosis. *In Vitro*, **30**, 53–8.

Pretlow, T. G. and Pretlow, T. P. (Series Editors) (1982–1984), *Cell Separation: Methods and Selected Applications*, Volumes 1–3. New York, Academic Press.

Purchase, I. F. H. and Conning, D. (Eds) (1986). International Conference on Practical *in vitro* Toxicology. *Food Chemical Toxical.*, **24**, 447–818.

Roth, K. S., Holtzapple, P., Genel, M. and Segal, S. (1979). Uptake of glycine by human kidney cortex. *Metabolism*, **18**, 677–81.

Ross, B. D. and Guder, W. G. (1982). Heterogeneity and compartmentation in the kidney. In *Metabolic Compartmentation*, Ed. Sies, H., pp. 363–409. London, Academic Press.

Rush, G. F., Smith, J. H., Newton, J. F. and Hook, J. B. (1984). Chemically induced nephrotoxicity: Role of metabolic activation. *CRC crit. Rev. Toxicol.*, **13**, 99–160.

Rylander-Yueh, L. A., Gandolfi, A. J. and Brendel, K. (1985). Use of isolated rabbit renal tubules to examine inhibition of organic acid/base transport by toxins. In *Renal Heterogeneity and Target Cell Toxicity*, Eds. Bach, P. H. and Lock, E. A., pp. 531–4. Chichester, J. Wiley.

Saier, M. H. (1981). Growth and differentiated properties of a kidney epithelial cell line (MDCK). *Am. J. Physiol.*, **240**, C106–C109.

Savin, V., Karniski, L., Cuppage, F., Hodges, G. and Chanko, A. (1985). Effect of gentamicin on isolated glomeruli and proximal tubules. *Lab. Invest.*, **52**, 93–102.

Scheinman, J. I. and Fish, A. (1978). Human glomerular cells in culture. Three subcultured cell types bearing glomerular antigens. *Am. J. Pathol.*, **92**, 125–39.

Scholer, D. W. and Edelman, I. S. (1979). Isolation of rat kidney cortical tubules enriched in proximal and distal segments. *Am. J. Physiol.*, **237**, F350–F359.

Schlondorff, D., Roczniak, S., Satriano, J. A. and Folkert, V. W. (1980). Prostaglandin synthesis by isolated rat glomeruli: effect of angiotensin II. *Am. J. Physiol.*, **239**, F486–F495.

Sraer, J., Foidart, J., Chansel, D., Mahieu, P. and Ardaillou, R. (1980). Prostaglandin synthesis by rat isolated glomeruli and glomerular cultured cells. *Int. J. Biochem.*, **12**, 203–7.

Sraer, J., Rigaud, M., Bens, M., Rabinovitch, H. and Ardaillou, R. (1983). Metabolism of arachidonic acid via the lipoxygenase pathway in human and murine glomeruli. *J. Biol. Chem.*, **258**, 4325–30.

Stollenwerk Petrulis, A., Aikawa, M. and Dunn, M. J. (1981). Prostaglandin and thromboxane synthesis by rat glomerular epithelial cells. *Kidney int.*, **20**, 469–74.

Striker, G. E., Killen, P. D. and Farin, F. M. (1980). Human glomerular cells *in vitro*: Isolation and characterization. *Transplant Proc.*, **12**, 88–99.

Stumpf, B. and Kraus, H. (1978). Inhibition of gluconeogenesis in isolated rat kidney tubules by branched chain alpha-ketoacids. *Pediat. Res.*, **12**, 1039–44.

Taub, M. and Saier, M. H. (1981). Amiloride-resistant madindarby canine kidney (MDCK) cells exhibit decreased cation transport. *J. Cell. Physiol.*, **106**, 191–9.

Trifillis, A. L., Regec, A. L., Hall-Craggs, M. and Trump, B. F. (1985a). Effects of cyclosporin on cultured human renal tubular cells. In *Renal Heterogeneity and Target Cell Toxicity*, Eds Bach, P. H. and Lock, E. A., pp. 545–8. Chichester, J. Wiley.

Trifillis, A. L., Regec, A. L. and Trump, B. F. (1985b). Isolation, culture and characterization of human renal tubular cells. *J. Urol.*, **133**, 324–9.

Turner, P. (Ed.) (1983). *Animals in Scientific Research: An Effective Substitute for Man?* London, Macmillan Press.

Udenfriend, S. (1969). *Fluorescence Assay in Biology and Medicine*. New York, Academic Press.

Viano, I., Eandi, M. and Santiano, M. (1983). Toxic effects of some antibiotics on rabbit kidney cells. *Int. J. Tiss. React.*, **5**, 181–6.

Weil, C. S. (1972). Guideline for experiments to predict the safety of materials for man. *Toxicol. appl. Pharmacol.*, **21**, 194–9.

Williams, G. M., Dunkel, V. C. and Ray, V. A. (Eds) (1983). Cellular systems for toxicity testing. *Ann. N.Y. Acad. Sci.*, **407**, 1–482.

Zenser, T. V., Mattammal, M. B., Brown, W. W. and Davis, B. B. (1979). Cooxygenation by prostaglandin cyclooxygenase from rabbit inner medulla. *Kidney int.*, **16**, 688–94.

Zimmermann, U. (1982). Electric field-mediated fusion and related electric phenomena. *Biochim. Biophys. Acta*, **694**, 227–77.

Zimmermann, U., Scheurich, P., Pilwat, G. and Benz, R. (1981). Cells with manipulated functions: New perspective for cell biology, medicine and technology. *Angew. Chem.* (*Int. Ed. Engl.*), **20**, 325–44.

CHAPTER 8

Testing chemicals for acute toxicity and irritancy: the contribution of *in vitro* techniques

C. Rhodes, G. J. A. Oliver, M. A. Pemberton and R. C. Scott

Introduction

For hundreds of years the assessment of acute health effects, such as acute systemic poisoning and cutaneous and ocular toxicity, following intentional or adventitious exposure to chemicals has been important to their use by man. During the early part of the twentieth century such studies became more formalized. Since the 1940s, various protocols have been used to detect systemic toxicity following oral ingestion, inhalation and dermal contact, as well as local effects, such as contact sensitization and dermal and ocular irritation, which can occur following exposure of the external tissues to a chemical. In later years, the development of *in vitro* and *ex vivo* techniques using cultured cells, tissues and organs to make an assessment of toxic responses has been pre-eminent. Perhaps most visible to the general scientific community has been the development of *in vitro* genotoxicity screens such as the Ames *Salmonella* mutation assay for the assessment of potential mutagenicity and carcinogenicity. However, considerable effort has gone into *in vitro* and *ex vivo* procedures for the assessment of other adverse effects of chemicals (King *et al.*, 1986; Purchase *et al.*, 1986), such as the development of pre-screens to evaluate skin corrosive potential (Oliver *et al.*, 1986). Whilst these techniques can detect the potential for an adverse effect, there is still much debate on whether *in vitro* and *ex vivo* techniques can provide sufficient evidence to demonstrate the presence or absence of an adverse effect without recourse to *in vivo* studies. Toxicology studies are mainly concerned with the detection of adverse effects but they often extend to the further characterization of observed effects and less often to the determination of the biochemical mechanism underlying the adverse change.

Acute systemic toxicity studies are primarily concerned with the detection of adverse changes to the body systems, e.g. the central nervous system, cardiovascular system, respiratory system, excretory system, which are likely to be life-threatening following single exposure to a chemical. Additionally, acute toxicity studies are also concerned with non-life threatening responses that occur to the external tissues but which may lead to some loss of function or ill health and a reduction in quality of life, e.g. dermal and ocular toxicity. Clearly, there is a need to distinguish those substances that are immediately harmful to man (and animals) from those that are not. This, in our view, is the primary purpose of acute toxicity studies in the chemical industry. Currently acute toxicity studies are not intended to provide, and cannot be expected to do so, a deeper understanding of the biochemical mechanisms of toxicity. Current protocols, using animals, allow the toxicologist to identify those substances for which extra care is needed, or those which are too toxic for the proposed use.

Whilst our society demands the safeguard of human health and the environment as prerequisites for the development, manufacture and use of chemical substances, there is also a quite natural concern for the welfare and humane treatment of laboratory animals used to provide the necessary information on clinical toxicity.

Our purpose in this chapter is to discuss the use of *in vitro* techniques, for the assessment of acute toxicity. The work described has arisen from the efforts of a number of scientists in the ICI Laboratories at Alderley Park and forms a part of a major programme with objectives to assess, develop and introduce *in vitro* techniques into pharmacology and toxicology.

Acute toxicity studies *in vivo*

It is often forgotten that within toxicology there has been a continuing and serious debate on the most appropriate method for 'the quantitative study of injurious effects of chemical and physical agents as observed in the alteration of the structure, function and response of living systems, including the evaluation of safety'. The phrase in quotation marks represents the definition of toxicology preferred by the European Society of Toxicology (Holmstedt, 1976). It is clear that acute toxicity studies in particular have achieved a level of notoriety in the public domain due to the efforts of animal welfare groups. The primary focus has been the 'LD_{50} test' and the 'Draize irritancy test'. It has been stated openly that this was a simple application of the Archimedean principle of leverage, by which (to move anything) a lever and fulcrum is used to transform a small amount of applied effort into a much greater force (Spiro, 1985). The fulcrum or point of leverage has been the 'LD_{50}' and 'Draize tests' with *in vitro* alternatives the lever to which the effort has been applied.

Unfortunately a considerable amount of rhetoric has been used in the description of these tests and of the toxicologists who use them, which often overshadows more reasoned debate.

Acute toxicity studies are only a small proportion of the spectrum of laboratory studies for which animals are used (Table 8.1). They are a key component in any industrial laboratory's assessment of health effects of chemicals. A review has been made of the studies completed within the Central Toxicology Laboratory of ICI to evaluate systemic toxicity, cutaneous and ocular irritancy and contact sensitization (Table 8.2) against a broad spectrum of chemical types (Table 8.3). But what is also important to note is that the majority of such studies did not produce significant evidence of toxicity (Fig. 8.1). Therefore, the use of *in vitro* procedures has to be considered against this background and a strategy for their use developed. Within our own laboratory this strategy has been targetted primarily on the development of *in vitro* techniques for use as *in vitro* pre-screens to *in vivo* studies and not necessarily as replacements (Fig. 8.2). The application of such a strategy of tier testing to eliminate the relatively severe responses does establish a reduction in numbers of animals and introduces possibilities for refinement of *in vivo* studies, as well as

Table 8.1 Home Office UK statistics of experiments on living animals*

Parameter	*1980*	*1985*
Total number of experimental animals used	4 579 478	3 280 135
Acute toxicity		
all purposes	10.59%	11.05%
other than medical, veterinary or dental	4.11%	5.00%
industrial substances	1.10%	1.06%

*Statistics of experiments on living animals (1980, 1985). Her Majesty's Stationery Office, London.

Table 8.2 Proportion of study types in data base

Study type	*Proportion* (%)
Systemic	
oral	20
dermal	11
Irritation	
skin	26
eye	22
Allergenicity	
contact sensitization	21

minimizing discomfort to those animals which are used to confirm the 'absence of acute health effects'.

Acute systemic toxicity in vitro

CELL CULTURE

The majority of us are generally aware that animals and plants are complex biological entities, often consisting of several organs, thousands of enzymes, tens of thousands of proteins, hundreds of thousands of different molecules and millions of cells. From birth to death, the fertilized human cell goes through 10^{16} cell divisions. The diversity of animal species is built on an increasing level of organization from chemicals through cells, through tissues to organs, which make up the principal physiological systems of the organism itself (Table 8.4). It is now well established that all animal cells have the same basic features of a cell membrane enclosing sub-cellular organelles for cellular regeneration, reproduction,

Table 8.3 Analysis by chemical type

Chemical type	*Proportion* (%)
Pesticides	44
Intermediates, dyestuffs, and speciality chemicals	20
Pharmaceutical intermediates	11
Cutting Oils, biocides	11
Chlorinated solvents, surfactants	10
Miscellaneous formulations	4

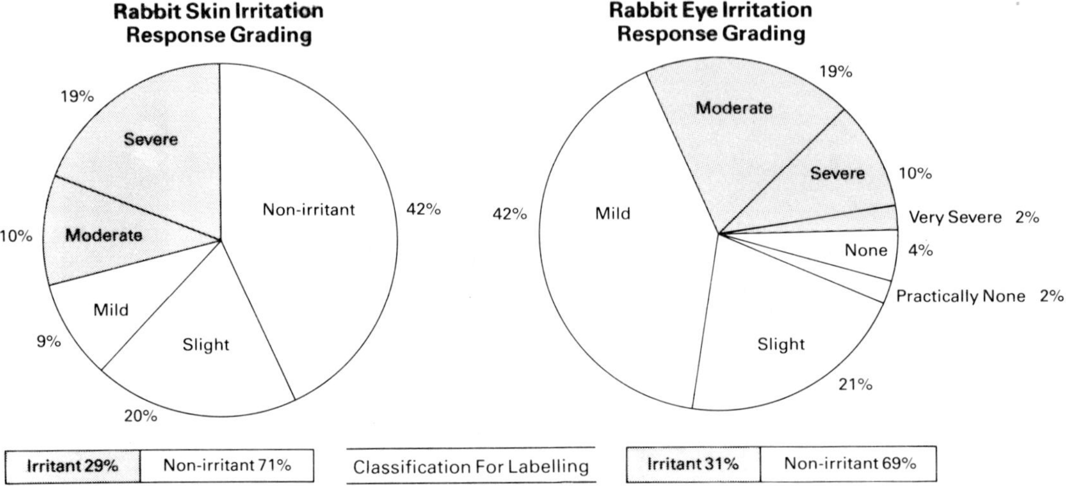

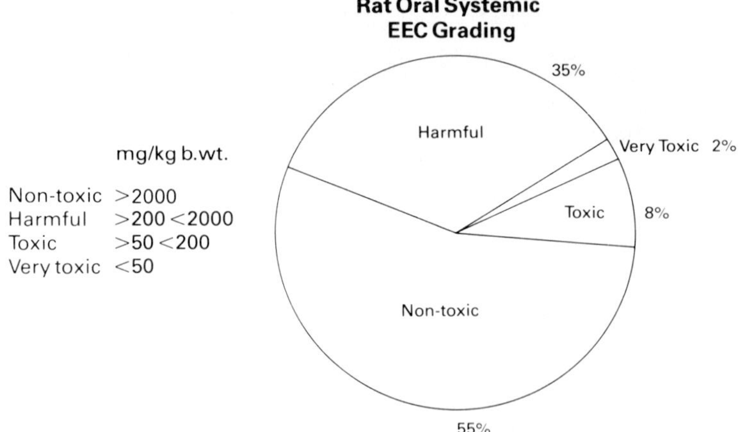

Fig. 8.1 Prevalence of response in acute systemic toxicity, skin and eye irritancy studies

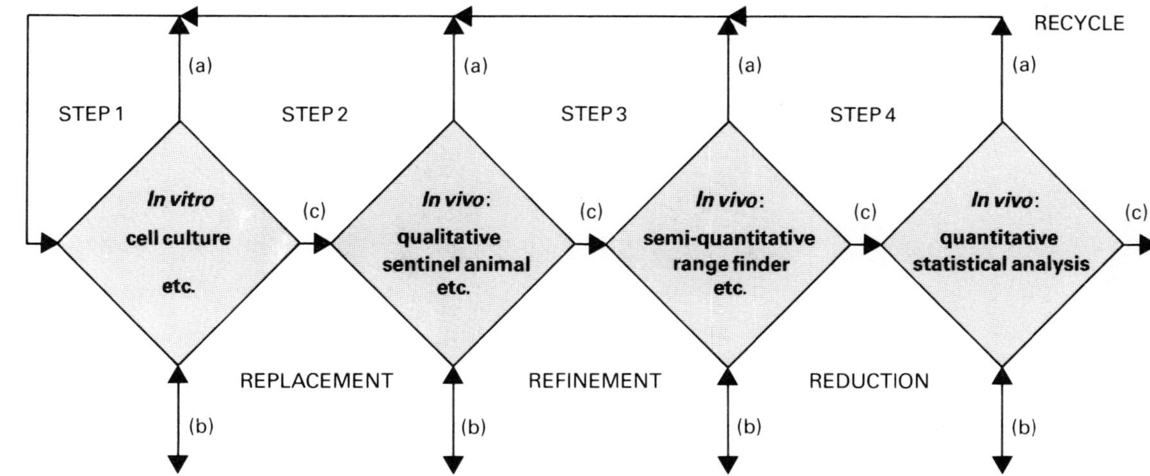

(a) Insufficient or inadequate data. Re-evaluate, go back to previous step.

(b) Data adequate to establish toxicity categorisation for intended use of chemical.

(c) Data adequate to establish potential for toxicity but inadequate to establish toxicity categorisation for intended use of chemical.

Fig. 8.2 Step sequence approach to the use of *in vitro* techniques as pre-screens to the *in vivo* assessment of acute toxicity

Table 8.4 Physiological systems of mammalian organisms

System	*Organs/tissues*
Integumentory	Skin, fingernails
Skeletal	Bones, cartilage
Muscular	Muscles, tendons
Nervous	Brain, spinal cord, nerves, sense organs (eye, ear)
Endocrine	Testis, ovaries, small intestine, pancreas, stomach, kidney, adrenals, thymus, thyroid, parathyroid, pituitary, pineal
Cardiovascular	Heart, spleen, blood vessels
Lymphatic	Lymph, lymph nodes, lymph vessels, lymph glands, spleen, thymus, tonsils
Respiratory	Lungs, nasal cavity, oral cavity, pharynx, larynx, trachea, bronchus
Digestive	Oral cavity, salivary glands, pharynx, oesophagus, stomach, pancreas, small intestine, large intestine, rectum, colon, gall bladder, liver
Urinary	Kidney, ureter, urinary bladder, urethra
Reproductive	Prostate gland, vas deferens, penis, testis, uterine tube, ovary, uterus, vagina

and energy generation as well as other more specialized cellular functions. A fundamental assumption in the *in vitro* assessment of systemic toxicity is that the disturbance of any part of the basic cellular system by toxins should lead to a common response such as a general cytotoxicity (Benson *et al.*, 1986). Many different cell types have been used in cytotoxicity investigations and a large number of endpoints for detecting cytotoxicity have been used, e.g. the determination of cell viability, cell morphology, cell proliferation, membrane damage, uptake or incorporation of radioactive precursors and metabolic activity. Probably hundreds of chemicals have now been studied in *in vitro* systems to determine the cytotoxic response of the cells to various doses of the chemical. Many investigators have claimed that tests of general cytotoxicity, in a number of different cell types, give comparative ranking of the cytotoxic potential for a variety of chemicals, but this is open to some dispute (*ECETOX Monograph*, 1985).

On reading the published literature it is difficult to perceive exactly how the information from cell cytotoxicity tests can be used. One possibility may be the development of a classification scheme for transforming the cytotoxicity ID_{50} values into categories of potential hazard statements comparable to those used for LD_{50} values (Table 8.5). For some industrial chemicals which are not to be manufactured in large amounts and are not for intentional exposure to man, then the ability to label the potential hazard from a cytotoxicity index may well be all that is required, e.g. $ID_{50} < 10\,\mu g/ml$ potentially very toxic, >10 but $<100\,\mu g/ml$ potentially toxic; >100 but $<1000\,\mu g/ml$ potentially harmful; $>1000\,\mu g/ml$ potentially non-toxic. Obviously, there are many assumptions being made in this direct extrapolation from single cells to complex integrated organisms. There are many non-toxicologists who would not be able to recognize the limitation of a

Table 8.5 Criteria for classification of chemicals on the basis of LD_{50} values from acute oral systemic toxicity data for solids

Country/organization	*Scheme**			
USA				
Super toxic (<5)	Highly toxic (<50)	Very toxic (<500)	Moderately toxic (<5000)	Slightly toxic (<15 000)
Switzerland				
Category 1 (<5)	Category 2 (<50)	Category 3 (<500)	Category 4 (<5000)	Category 5 (<15 000)
World Health Organization				
Extremely hazardous (<5)	Highly hazardous (<50)	Moderately toxic (<500)	Slightly toxic (<5000)	
United Nations				
Toxic 1 (<5)	Toxic 2 (<50)	Toxic 3 (<500)		
European Economic Community				
Very toxic (<25)	Toxic (<200)	Harmful (<2000)		
United Kingdom/Pesticide Safety Precaution Scheme				
Very toxic (<5)	Toxic (<50)	Harmful (<500)		
Japan				
Toxic (<30)	Deleterious (<300)			

*The classification category (e.g. Toxic 1) is followed by the LD_{50} values (in parenthesis) in mg/kg bodyweight which define the limit for a category.

potential hazard label derived from a cytotoxicity test compared with a hazard label derived from an *in vivo* study using another mammal as a surrogate for man.

What separates the ranking of chemicals by an *in vitro* cytotoxicity screen from those ranked by a whole organism system is the response of specialized functions of selected cells and tissues within the whole organism. It is these very specialized cells and tissues which not only maintain homeostasis but also govern how much, and at what rate, a chemical is absorbed, transformed, distributed to the tissue and eventually excreted. The determination of the lethal dose in a rat provides data which can rank chemicals with regard to their potential to cause lethality but may not necessarily predict the human toxic dose.

Similarly, *in vitro* cytotoxicity studies provide data which allow some ranking of the potential for chemicals to cause toxicity, but cannot predict whether the effect will occur at all, or the dose at which toxicity will be evident in a whole organism. The qualitative and quantitative pharmacokinetic and metabolic dissimilarities between species which compromise the extrapolation from rat to man, compromise much more the extrapolation from *in vitro* cellular systems to man even when human tissues and cells are used.

With our present knowledge of toxic mechanisms, pharmacokinetics and pharmacodynamics, *in vitro* cell cytotoxicity screens cannot replace studies *in vivo*. They may, as our knowledge develops and in combination with well-designed computer models, lead eventually to the prediction of a chemical's toxic potential. However, their current application in the assessment of toxicity is limited to well-defined circumstances. *In vitro* systems are of limited value as general pre-screens to determine the intrinsic potential for a chemical to exert toxicity *in vivo*. For some purposes and use of chemicals this may be sufficient without further recourse to *in vivo* studies, e.g. pyrogenicity evaluation using the *Limulus* amoebocyte lysate test.

There are *in vitro* techniques other than cytotoxicity assays which can contribute to the *in vivo* assessment of systemic toxicity (Gillette, 1986). Within our own laboratory we have focused on monitoring the absorption of chemicals using *in vitro* techniques, and like many others we have developed *in vitro* systems to study biotransformation.

IN VITRO ABSORPTION

Skin permeability

With the majority of industrial chemicals, dermal contact is often the most significant route of exposure during manufacture, formulation or use of products. Knowledge of the rate of skin permeability of a chemical can make a valuable contribution to the prediction of the risk of an adverse effect following exposure. The diffusion of chemicals through the skin is a passive process in which penetration of the stratum corneum (the outermost layer consisting of dead cornified epidermal cells) is the rate-limiting step (Fig. 8.3a). The underlying epidermis and dermis do not contribute significantly in determining the absorption rate of a chemical. As a consequence, it is not necessary to keep skin viable in order to determine the absorption rate, and this has led to the development of an *in vitro* technique for the determination of the percutaneous absorption of chemicals through excised skin samples from animal and human cadavers (Dugard, 1986; Scott *et al.*, 1986a). The technique is simple and employs a glass diffusion cell (Fig. 8.4) which consists of a donor chamber separated by a skin membrane from a receptor chamber. The penetrant chemical is applied to the epidermal surface and the lower chamber is filled with a receptor fluid. Periodically, the receptor fluid is analysed to determine the amount of chemical and an absorption rate derived.

This technique is now used routinely within the Central Toxicology Laboratory (ICI, Macclesfield) to compare the percutaneous absorption of chemicals through human and rat skin (Dugard *et al.*, 1983; Scott *et al.*, 1986b). For example, esters of phthalic acid may contact the skin during their manufacture and use as plasticizers. The *in vitro* percutaneous absorption rate through rat skin (the species used for *in vivo* toxicity assessment) of several esters was compared with their percutaneous absorption rate through human skin (Table 8.6). Rat skin was considerably more permeable than human skin. It is possible that *in vitro* permeability studies may eventually be used as pre-assessments prior to *in vivo* systemic toxicity studies. However, it must be remembered that the determination of *in vitro*

SKIN IRRITATION

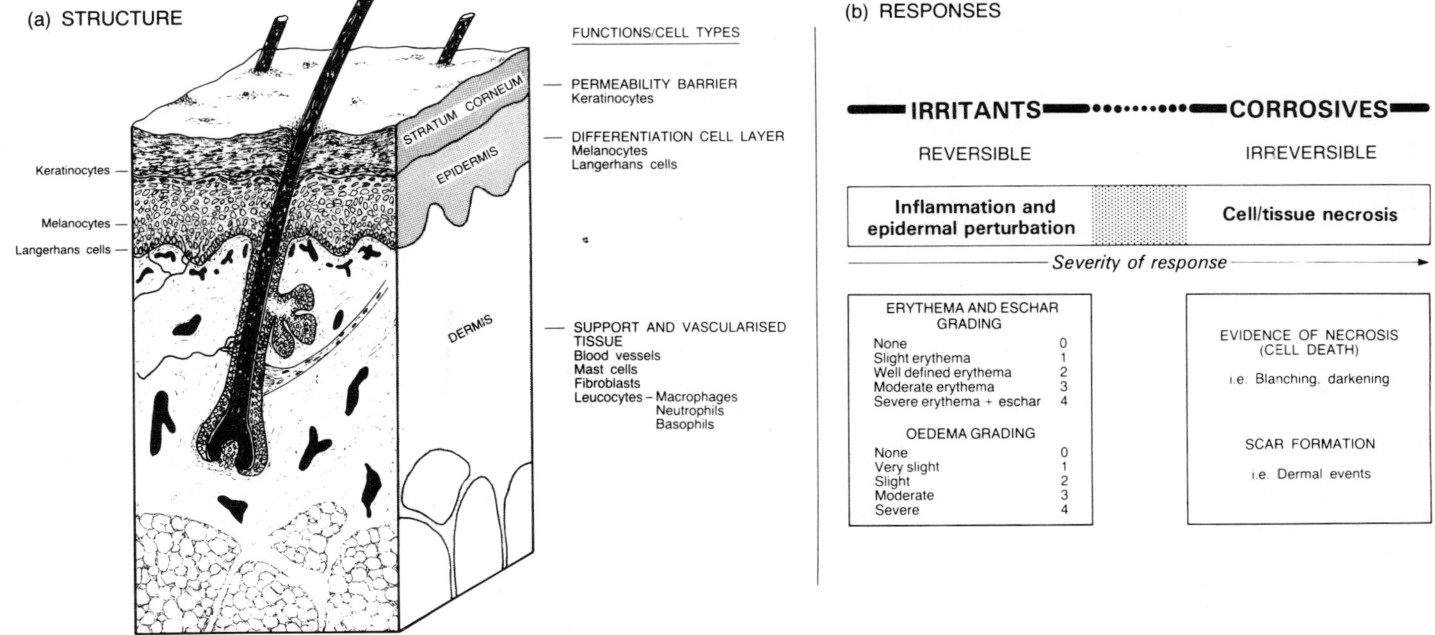

Fig. 8.3 (a) The structure of the skin and (b) the grading of responses to determine skin irritancy

absorption rates requires the quantitative analysis of the chemical under study. This can be achieved either by radioactively labelling the chemical or by the development of an appropriate analytical procedure using chromatographic techniques. For many industrial substances the development of analytical techniques is appropriate for single materials but not for complex mixtures.

Similar *in vitro* techniques can, and are, being applied to the assessment of absorption from the gastrointestinal tract although this requires the tissue to be viable and is therefore more technically difficult.

Gastrointestinal absorption

The gastrointestinal barrier that separates the gastrointestinal (GI) lumen and the splanchnic blood and lymphocyte circulation is, like the skin, a complex tissue. The permeability characteristics of the GI tract have been studied for many types of chemicals including nutrients, drugs and industrial chemicals (Rozmann and Hanninen, 1986). In general, the rate and extent (bioavailability) of absorption of a substance is determined by comparison of the pharmacokinetic analysis of the concentration and amount of the substance and metabolites in samples of blood, tissue, urine and other excreta taken during *in vivo* studies on animals or man following oral and parenteral exposure of the substance. *In vivo* bioavailability data do not often provide much information on the mechanisms of absorption, whereas a variety of *in vitro* and *in situ* techniques permit the direct assessment of the permeability characteristics in an isolated intestinal segment. An *in vitro* method for investigating intestinal absorption involves the isolation of a small segment of the gastrointestinal tract, to place the mucosal surface to the outside and the serosal to the inside.

After filling the lumen with physiological medium, tying off each end of the segment,

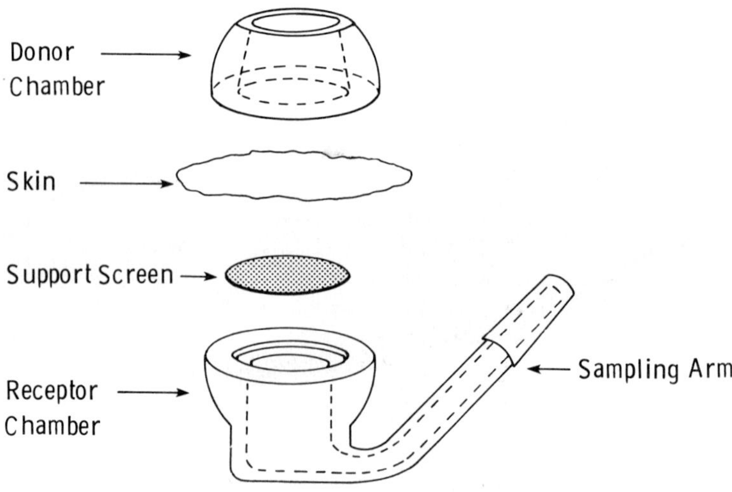

Fig. 8.4 Percutaneous absorption. Horizontal diffusion cell for determining *in vitro* absorption through cadaver transcutaneous skin and epidermal membranes

Table 8.6 *In vitro* percutaneous absorption of Phthalate esters through rat and human epidermal membranes

Compound	*Absorption rate* ($\mu g/cm^2/h$)	
	Human	*Rat*
Dimethyl phthalate	4	42
Diethyl phthalate	1	41
Di-*n*-butyl phthalate	<1	9

and placing the intestinal sac in a flask of physiological saline, the permeation of substance from the bathing fluid into the lumen of the sac can be monitored (Wilson and Wiseman, 1954) and with the placement of a glass cannula at each end serial sampling of the lumen content can be achieved (Crane and Wilson, 1958). A limitation of this procedure is that it is avascular (without an intact blood circulation). In the *in vitro* technique a substance has to permeate the entire thickness of intestinal wall from mucosal to serosal surface through the epithelium, submucosa, muscularis and serosa whereas with a functional vasculature the compound has only to traverse the epithelial layer between the mucosal surface and the villous blood capillaries (Perrier and Gibaldi, 1973). Currently *in situ* methods in anaesthetized animals are the only methods by which segments of the gut can be isolated and perfused with chemical whilst maintaining the blood circulation intact (Doluisio *et al.*, 1969).

Biotransformation

In vitro techniques to monitor biotransformation have been reviewed at a previous symposium of the Humane Research Trust (Fry, 1983). Whilst isolated cell systems have been developed within the Central Toxicology Laboratory in Macclesfield, these are not routinely applied to the assessment of acute systemic toxicity but are used in more specific evaluations of the biochemical mechanisms of toxicity (Lhugenot *et al.*, 1985), and therefore they will not be further discussed. However, it would seem axiomatic that in order to establish any credibility for *in vitro* cytotoxicity assays to predict potential toxicity *in vivo*, then *in vitro* assessment of absorption and biotransformation needs to be part of any basic strategy.

Primary cutaneous toxicity

An excellent review relating to the *in vitro* prediction of cutaneous toxicity has recently been published (Parish, 1986). The features of cutaneous irritation and corrosivity which are applicable to *in vitro* tests using organ or cell cultures are the initial damage to the keratinocytes, epidermal cells, leukocytes and other inflammatory cells which contribute to subsequent tissue damage. The secondary effects, which relate to cellular migration from the circulating blood, currently cannot be readily examined by isolated tissues and cell systems. Imaginative experimental design may yet lead to appropriate developments.

Since skin is the point of entry of many substances into the body and because injury to the skin may occur in the locality of chemical contact, knowledge of cutaneous toxicity is important for hazard assessment. The skin, like many other organs is complex in both structure and function (Fig. 8.3). Substances that interact with this tissue can produce differing toxic effects. A number of *in vitro* and *ex vivo* systems have been developed which may be effective as pre-evaluations to identify chemicals which can penetrate the stratum corneum and which may then produce cutaneous toxicity.

The approaches that have been made are influenced by the fact that the skin can be visualized into three distinct parts: The stratum corneum, a layer of dead compact cells (keratinocytes) forms the outermost layer of the skin. It acts as a barrier to the penetration of substances both into and out of the body. The underlying epidermis consists of layers of viable cells, originating from a basal layer, which differentiate to form keratinocytes that are incorporated into the stratum corneum. Highly specialized cells can be found in the epidermis, e.g. the melanocytes and Langerhans cells (the so-called antigen presenting cells). The dermis forms a fibrous, elastic support for the epidermis and contains a sparse population of fibroblasts. Blood vessels (and associated mast cells) permeate the dermis in which various skin appendages (hair follicles, sebaceous glands) also occur. With such a heterogeneous tissue it is clear that cutaneous toxicity may be mediated by complex mechanisms which may include the rate of chemical penetration and subsequent effects such as denaturation of proteins, reduction of barrier function, increase in vascular permeability (mediated by histamine release), cytotoxicity, chemotaxis, inhibition of DNA, and/or protein synthesis and leakage of degradative intracellular enzymes. The material may cause inflammation by being directly irritant, by evoking a cell-mediated delayed hypersensitivity response (allergic contact dermatitis) or by interacting with light energy to cause a photocontact dermatitis. Other types of reaction to the skin have been described, including the sensation of stinging or burning without visible skin change (paraesthesia) to severe erosion of the skin (corrosivity), itching (contact urticaria), depigmentation, hyperpigmentation and follicular reactions with the development of acne-type lesions. Whereas *in vitro* techniques are limited and will be targetted to some specific change, whole animal studies do have the advantage of being able to detect a range of such effects by direct observation.

ORGAN CULTURE

Skin corrosivity

The first step from an *in vivo* study to an *in vitro* technique is the use of isolated organs. This has been the focus of much of the work in the ICI laboratories with regard to skin corrosivity. An analysis of the *in vivo* skin response to a number of corrosive and irritant chemicals indicated that corrosive agents exert a direct and relatively rapid physiochemical effect on the tissue. Such an effect results in direct degradation of the stratum corneum impairing its normal function as a permeability barrier. Earlier studies of skin permeability had shown that the electrical resistance of the skin changed if the skin permeability barrier was damaged (Fig. 8.5). The chemically-induced reduction of stratum corneum integrity has been measured using a change in electrical resistance across the skin slice (Oliver *et al.*, 1986, 1988). The epidermal slice resistance model has been extensively validated as a technique for the *in vitro* prediction of skin corrosive potential. Keratome slices of skin containing stratum corneum and a full depth keratinocyte layer (epidermis) were mounted at one end of a PTFE tube enabling chemicals to be applied to the corneal surface with the dermis exposed to the supporting medium (Fig. 8.6). The skin resistance of the epidermal slice was measured at varying times after exposure to test chemicals. By varying the contact time and the period between contact and measurement of resistance it was possible to achieve both high specificity (the proportion of non-corrosive chemicals giving a negative result) and high sensitivity (the proportion of corrosive chemicals giving a positive result). The technique is considered to have significant potential as a pre-screen for the identification of skin corrosive properties prior to animal tests, and to offer distinct advantages over *in vivo* models of skin corrosivity, since the data are objective and quantifiable and provide additional qualitative information concerning the action of the chemicals. One additional benefit arising from this approach has been the ability to compare the response of human cadaver skin *in vitro* with that of animal skin *in*

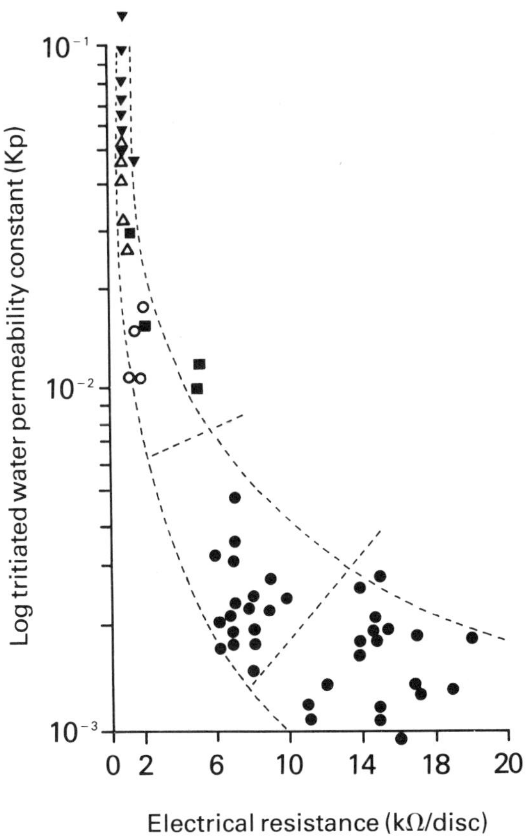

Fig. 8.5 The transcutaneous electrical resistance and tritiated water permeability of intact skin (●) and skin modified *in vitro* by slight abrasion (○), cellotape stripping (■), severe abrasion (△) or heat separation (▼)

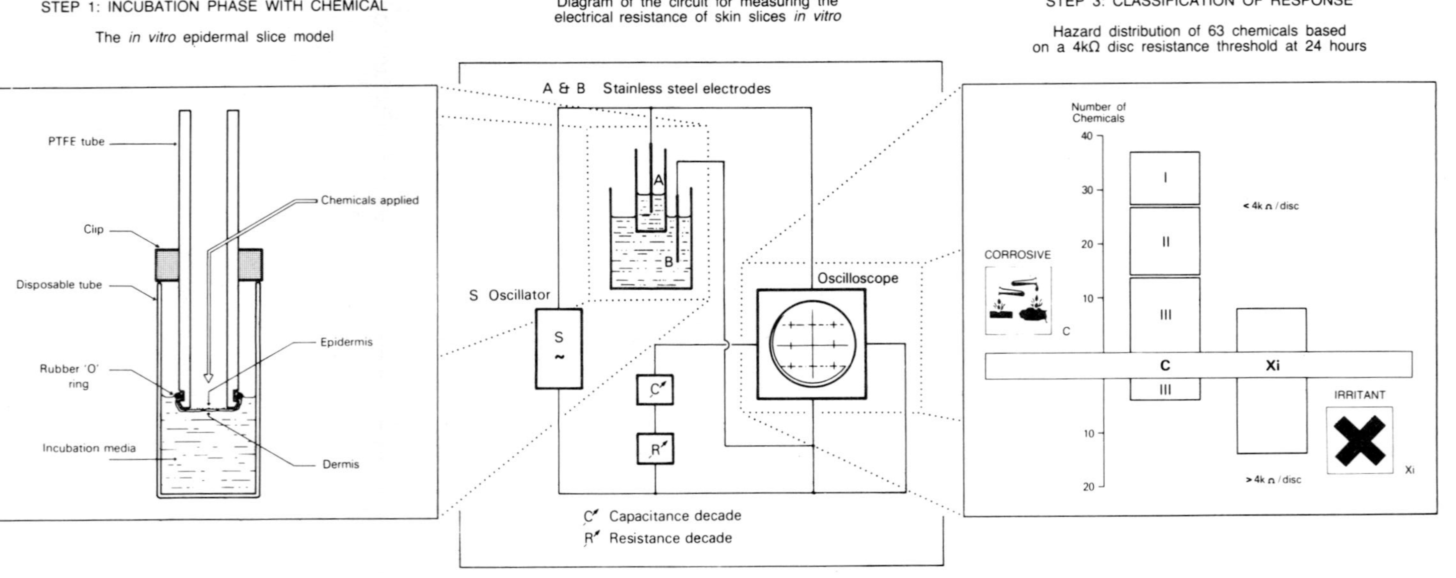

Fig. 8.6 The *ex vivo* epidermal skin technique for evaluation of corrosive potential

vitro and *in vivo* (Oliver and Pemberton, 1986). This adaptation may facilitate the use of human cadaver skin to predict potential skin corrosive hazard to man directly. Under similar conditions, only two-thirds of the chemicals that were corrosive in the *in vivo* rabbit studies and which were predicted as corrosive in the *in vitro* studies using rat skin were predicted to be corrosive in the *in vitro* studies with human skin. From these data human skin appears to be less susceptible to chemical corrosive action than does animal skin, and this may have implications with respect to the accurate assessment of human hazard.

Skin irritancy

More often slices of skin have been used to study the cytotoxic action of an irritant on epithelial/epidermal cells by monitoring, for example, the release of enzymes, histochemical change, increased or decreased utilization of amino-acids or nucleosides.

Unlike the epidermal resistance model described above, these techniques require viable tissue. This integrated system consisting of the epidermis and some dermis can be argued to be a more complete model of the skin than isolated cell systems. An applied compound has first to penetrate the stratum corneum before it can manifest a toxic response on the underlying epidermal cells. However, a functioning vasculature is absent from the dermis so that erythema and oedema, which may be accompanied by fibrin deposition and the infiltration of platelets, leucocytes and neutrophils cannot be observed. Clearly, the system allows the investigation of epidermal cells beneath the protective barrier of the stratum corneum, focusing on the direct cytotoxic effect rather than the subsequent inflammatory cell infiltration process. Such changes encompass the stimulation of metabolism, cell migration, mitosis, hyperplasia, hydropic degeneration (vacuolation), cell enzyme activation, chromatin aggregation, cell swelling, nuclear pyknosis, perinuclear vacuolation, autolysis and disappearance of enzymes activity.

The major difficulty appears to be the interpretation of changes occurring when broad, complex classes of chemical are used. Acidic substances tend to 'fix' the cells, whereas alkalis, dependent on their nature and concentration, tend to remove the lipid and dehydrate, and surfactants may loosen the cells and remove lipid. Failure to detect enzyme activity can be due to many factors, i.e. the destruction of enzymes within killed cells, fixation of the cell membranes so preventing enzyme release, and denaturation of the released enzymes.

The basic working hypothesis is that cytotoxicity of an organ culture of skin (essentially epidermal cells) measured *in vitro* may correlate with the *in vivo* responses which classify a chemical as an irritant. It is important to partition the overall *in vivo* irritation response into relevant components (type of response, severity ratings, initial and maximal erythema and oedema scores), to allow more extensive cross correlation with *in vitro* parameters.

We have established a database of different types of industrial chemical representing a range of skin and eye irritation potential. A total of 60 irritants have been selected and the *in vivo* data fully reviewed and verified. Using viable rat epidermal slices it is hoped that irritation potential will correlate with some of the biochemical indices of cytotoxicity, as exemplified for a severe, moderate and non-irritant chemical (Fig. 8.7).

In order to improve the sensitivity of the *in vitro* organ culture model, the above studies have been extended to include rabbit skin. The rabbit is regarded as a more sensitive species than the rat with respect to skin irritation. Conditions for the organ culture of rabbit skin can be applied up to 48 h. The studies in rabbit tissue are in an early phase, but on the basis of biochemical indices of cytotoxicity, rabbit skin *in vitro* appears to be of comparable sensitivity to rat skin (Pemberton and Oliver, 1986). There may be some difficulties in extending such techniques to maintain viable organ culture of human skin.

CELL CULTURE TECHNIQUES

The next level down from organ culture (the skin slice model) is primary skin cell culture in the form of cell monolayers. Because penetration of the chemical through the stratum corneal barrier is an intrinsic component of the irritant response in the skin, it has been recommended that an artificial stratum corneum is used. A barrier of corneal squares mounted in an agar medium to reduce the direct exposure of the primary cell layer to the direct application of the chemical has been proposed (Parish, 1986). Additionally, inflammatory change is a dynamic process and test design will eventually need to be such as to allow the monitoring of progressive change in damaged cells and the relationship between different changes. Such procedures prepare the way to examine interrelationships

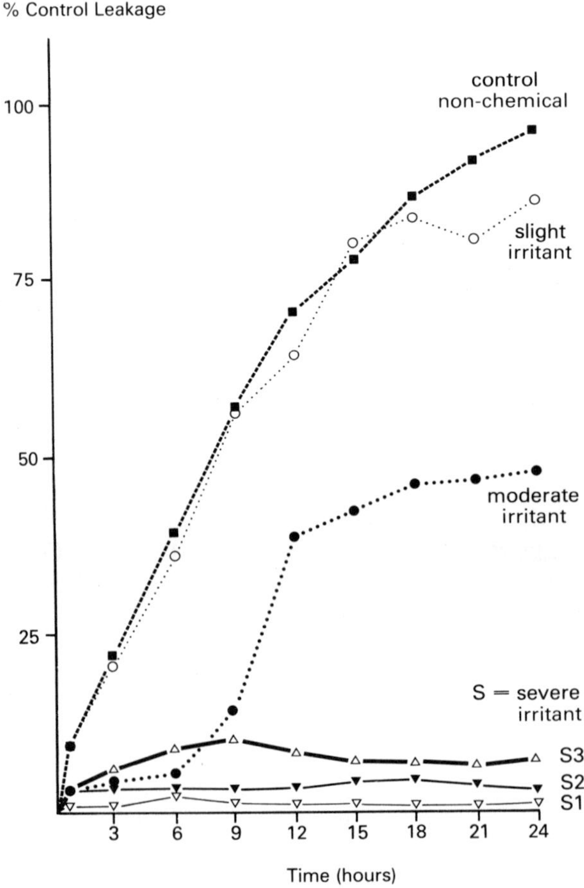

Figure 8.7 The effect of skin irritant chemicals on the accumulative leakage from intact young rat skin *in vitro*

between different types of cells and thus to detect indirect stimulation or damage to cells not directly exposed to a toxic substance. One important criterion of future model design will be the ability to include a stratum corneum (animal or human) between the cells and test material. The inclusion of a stratum corneum layer will regulate penetration of a chemical into the system in a manner similar to that *in vivo*. The kinetics of chemical delivery to epidermis and dermis contribute to species differences in response to irritant chemicals.

A single parameter assay system *in vitro* may be unsuitable for predicting toxicity of such a complex tissue as skin. Attempts have been made to predict irritancy using a combined assessment of cytotoxicity (reflected by cell proliferation, cell metabolism and cell membrane integrity) of 3T3 Swiss mouse fibroblasts, with an assessment of keratinocyte differentiation using a keratinising and stratifying epithelium cell line XB2 (Duffy *et al.*, 1986). Previously, other workers (Reinhardt *et al.*, 1985) had employed three cytotoxicity assays using baby hamster kidney fibroblasts, early and late human fibroblasts to measure cell detachment and cloning efficiency and growth inhibition under subculture conditions to assess 57 different chemicals. When the *in vitro* data were used to divide the chemicals into three classes: (1) non-irritant, (2) mild-to-moderate irritant or (3) strong irritant or corrosive, the authors claimed the *in vitro* data were more than 80% predictive of the classification derived from human exposure data *in vivo*.

However, the fact that skin slices and cell cultures are avascular is a significant limitation for studying the inflammatory response by such techniques *in vitro*. Immediate acute inflammation is manifested mainly by changes in the blood vessels, with erythema, oedema and fibrin deposition, and subsequent infiltration and activation of leucocytes. The chorioallantoic membrane of the embryonated chicken egg has been proposed as a model of an intact and functioning vasculature. Unfortunately, the response to injury is limited in that it has heterophils (chicken neutrophils), consequently it does not show mammalian-type inflammatory change. The membrane lacks the barrier afforded by the stratum corneum to the skin, which results in an all-or-nothing response according to the concentration of the chemical applied, with necrosis as the principal response, although evidence of inflammation with blood vessel changes was observed with some shampoo products (Parish, 1985). It would appear that superficially this model may provide the vascular component to the *in vitro* assessment of conjunctival but not cutaneous irritation. It must remain at this moment equivocal whether this model confers any advantage over avascular cell cultures. Currently, our laboratory has no experience of this technique. Others, however, remain unconvinced that the assessment used in this technique will stand up to a critical double-blind comparative study of a series of known irritants.

Ocular irritancy

It is important to stress that the eye, like the skin, is a complex organ (Fig. 8.8). No single test *in vitro* is likely to predict all toxic changes occurring within the specialized tissues that comprise a complex organ such as the eye. Macroscopic inflammatory responses of the eye are oedema, congestion and exudation of the conjunctivae. Such inflammatory change may show little tissue damage when examined histologically. The human cornea consists of an outer epithelium, between 5 and 10 cells deep, and is a non-vascular structure which, unlike the skin, has no protective cornefied cell layer (stratum corneum) over its surface. Because of this the cornea is subject to direct damage by applied chemicals as reflected in different degrees of opacity, such as translucent, opalescent and nacreous areas in the

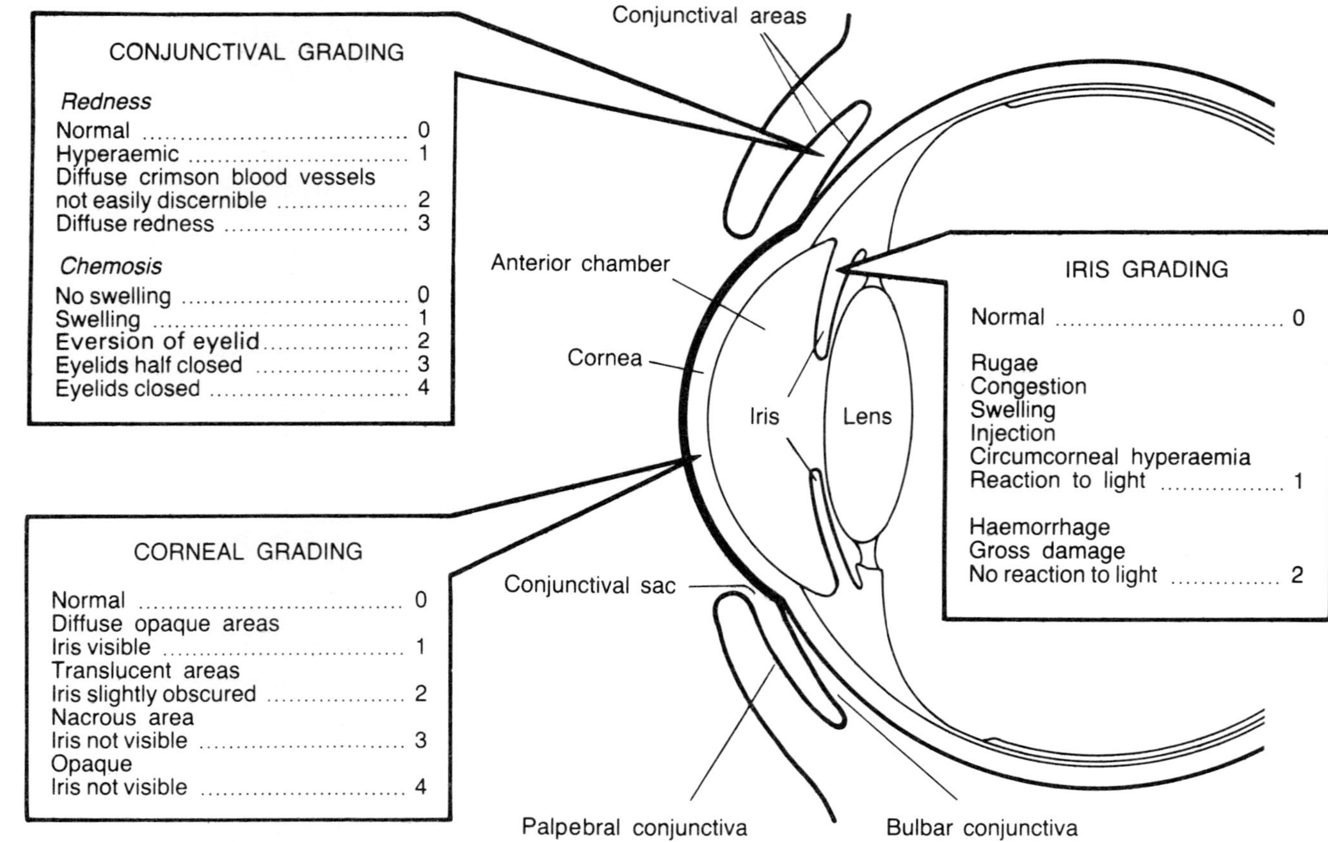

Fig. 8.8 Grading of *in vivo* ocular irritancy responses.

cornea. Consequently, cell cytotoxicity assessed by *in vitro* techniques may be indicative of corneal toxicity. Cell culture techniques purporting to reflect the direct damage to the conjunctival and corneal epithelium using monolayer and cell suspension techniques, have been extensively reported in the literature.

Within our own laboratory we are developing a model based on K562, human erythroleukaemic cells. Initial work examined the effects of a number of liquids with various levels of ocular irritancy determined previously from *in vivo* data. Both the dose–response and the time course for a fixed dose have been studied to standardize conditions for detecting a range of effects. Cell viability measured by trypan blue exclusion decreases linearly with increasing time of exposure to selected chemicals (Fig. 8.9). These preliminary results of the dose–response for 1 h exposure show a strong correlation between *in vitro* cytotoxicity, and severity ranking *in vivo* (Fig. 8.10), for this small group of ocular irritants. Such cell-culture techniques do provide a ranking of chemicals, but there are conflicting claims as to whether cytotoxicity should be used to classify chemicals as ocular irritants.

There are kinetic aspects which have to be considered with regard to application of chemicals to the intact eye, e.g. rapid removal from the corneal surface of applied solutions or fine chemicals as a consequence of the blink reflex, gravity and drainage, and the adherence of viscous materials.

The whole eye can be used as an isolated organ (Parish, 1985), and whilst a specialized perfusion apparatus has to be used, the technique overcomes many of the above constraints of *in vitro* cell culture techniques for the assessment of corneal damage. The chorioallantoic membrane of the embryonated chicken egg has been proposed for the assessment of ocular conjuctival irritancy and could thus complement the isolated organ procedure. There is still uncertainty whether the necrotic change of the chlorioallantoic membrane will adequately reflect the inflammatory and opacity changes induced by chemical irritants

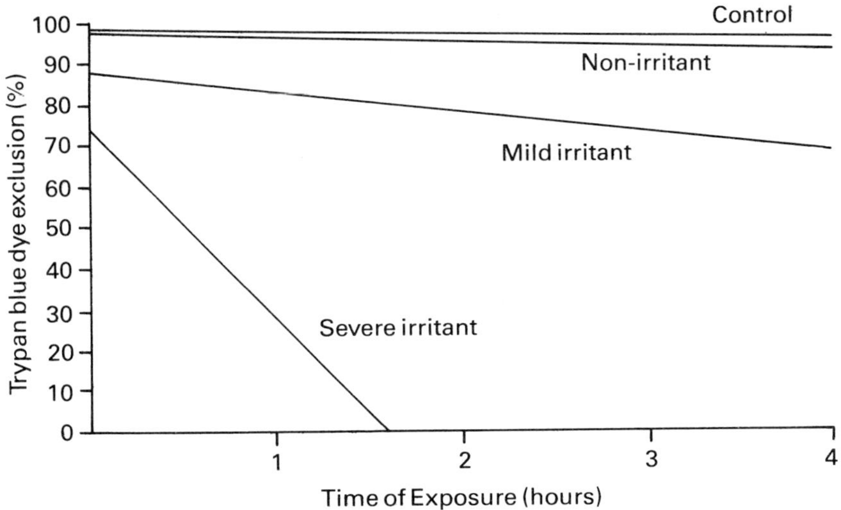

Fig. 8.9 Cell viability time course K562 human erythroleukaemic cells in the *in vitro* cell culture incubation system without stratum corneum in the presence of a non-irritant, a mild ocular irritant and a severe ocular irritant

(Lawrence *et al.*, 1986; Price *et al.*, 1986). The problem appears to be that the chorioallantoic membrane response is too sensitive, such that, of the 26 non-irritants examined by these workers, 15 showed positive results whereas only 1 of the 13 irritants failed to elicit a response. Therefore, whilst sensitivity is high (92%), specificity is low (42%).

In our view, techniques *in vitro* should be used as a pre-screen to identify the most potent cytotoxic agents such that subsequent studies *in vivo* can be designed to take this potency and potential irritancy into consideration by the use of dilution, anaesthesia, analgesics, etc. Eventually, through use of pre-screens, refinements may be achieved which could lead to the eventual acceptance of *in vitro* data as criteria for classification, although this does not appear to be feasible at the moment.

Allergic contact dermatitis

As well as primary cutaneous irritation (the immediate inflammatory response that may occur shortly after exposure to a chemical), a chemical may induce a local delayed hypersensitivity (cell-mediated immune response). This is an inflammatory cutaneous response occurring only in susceptible individuals who become immunologically sensitive and show a response only on subsequent exposure to a chemical. The complex molecular and cellular mechanisms of allergic contact dermatitis have only recently been more clearly defined (Fig. 8.11). There still remains considerable uncertainty, and further advances in knowledge will be required before *in vitro* techniques can be considered as replacements for *in vivo* procedures to test sensitizing chemicals (Parish, 1986). Parish has cogently stated: 'Immunologists apply *in vitro* techniques to study molecular and cellular interactions of sensitisation. The toxicologist needs to apply these studies to identify any allergic property

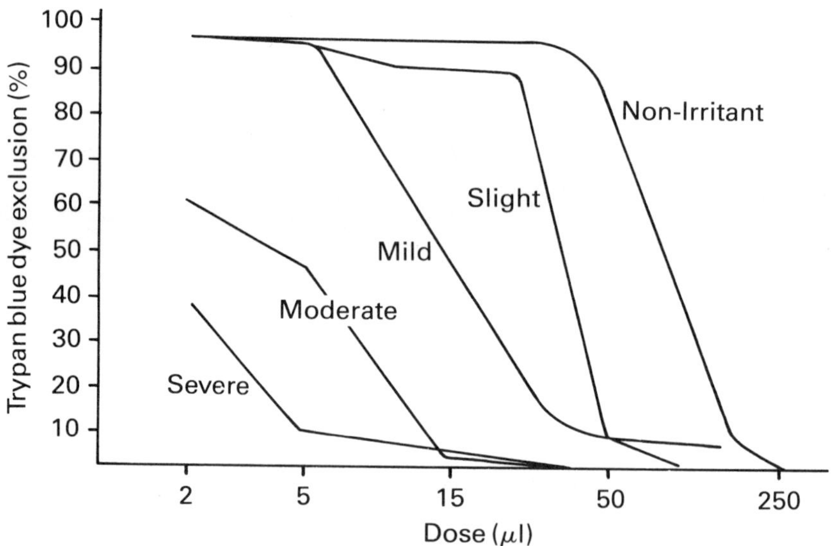

Fig. 8.10 Cell viability of K562 human erythroleukaemic cells following exposure for 1 h to selected liquids of known ocular irritancy *in vitro*

of a substance, predict its likely potency *in vivo* and ascertain the possible incidence, or risk of sensitization, in an exposed population'.

Immunologists in industrial toxicity laboratories are using *in vitro* and *in vivo* techniques to examine the complex mechanisms of allergic contact dermatitis (Kimber *et al.*, 1986; Botham *et al.*, 1987). Toxicologists are using various *in vivo* techniques, e.g. Stevens ear-flank test, the Beuhler flank test and the Magnusson and Kligman adjuvant maximization test, to detect contact hypersensitization potential. Apart from the determination of protein–chemical reactivity used to assess potential for respiratory sensitization (P. A.

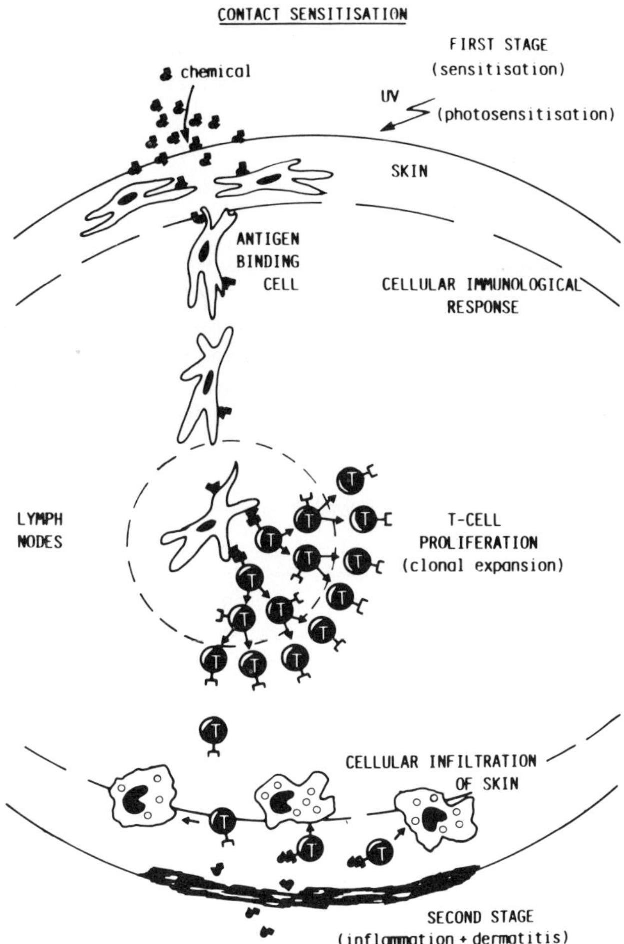

Fig. 8.11 Allergic contact dermatitis. Diagrammatic representation of the induction phase (demonstrating the absorption of chemical, chemical–Langerhans cell interaction, migration to lymph node and chemical–Langerhans cell complex stimulation of specific T-cell proliferation) and subsequent delayed hypersensitivity response (demonstrating the interaction of the specifically sensitized T-cell/lymphocyte with chemical or chemical–Langerhans cell complex to stimulate antigen non-specific lymphocytes) with consequent dermatitic (inflammatory) response)

Botham, personal communication), there have been few attempts to develop *in vitro* cell culture techniques to detect or predict the potential for chemicals to induce an allergic response. In our view, allergic contact sensitization is such a complex series of interactions, requiring clonal expansion of T-cells and the antagonist processes, which modulate the response, that it is essential to use integrated *in vivo* systems. Consequently, the focus of many studies has been the development of *in vivo* procedures with subsequent *ex vivo* techniques, which would allow objective and quantitative assessment of responses to replace the subjective observations of erythema and induration, which are confounded by primary irritation and coloured chemicals. A procedure has been developed using an evaluation of the proliferation of T-lymphocytes isolated from the draining lymph nodes of the mouse (Kimber *et al.*, 1986). This technique is currently going through a validation phase comparing its sensitivity and predictivity against the standard assays in guinea pigs. Such procedures will reduce the numbers of animals required to determine the sensitization potential of a chemical.

Summary and concluding comments

The impact of *in vitro* techniques on the number of animals used in acute toxicity studies will be achieved by the incorporation of validated models into a step-wise approach to toxicity assessment. Used as pre-screening tests, the information derived can be used to make a judgement regarding progress to the next (live animal) stage. This could result in the total avoidance of animal tests or the use of modified tests with fewer animals. The quality of the judgement on whether to proceed will depend on the individual test performance. The 'test performance' will depend on many factors, including test sensitivity, test specificity and the expected prevalence of positive effects (the proportion of chemicals having a particular type of toxicity within the test population of chemicals) (Rhodes *et al.*, 1987).

In vitro procedures will help in developing more humane approaches to acute toxicity studies but for the forseeable future they are unlikely to be accepted as the only assessment of acute toxicity and irritancy. In assessing the extent of a toxic hazard there are several important considerations related to the level of exposure, route of exposure, the severity of response, the speed of onset, duration and recovery from the effect. One important way of expressing the severity of response has been to establish the dose–effect or dose–response relationship. Such dose–effect (response) curves allow comparison and ranking of substances. The determination of the median effect (response) has allowed a convenient short-hand description of the potential ranking of the toxicity of a substance. Adverse responses such as corrosivity, irritancy, contact sensitization, are usually considered in a qualitative rather than quantitative fashion. Over many years, the toxic properties of substances, both the quantitative LD_{50} and the qualitative descriptors, have been derived from animal studies and published in the scientific literature, and extracted and quoted in such texts as *Sax's Dangerous Properties of Industrial Materials*, and the *Registry of Toxic Effects of Chemical Substances*. As such, they provide a rapid source of information of potential effects in mammals for those concerned with industrial and occupational health.

Various legislation and Health and Safety regulations place duties on the person who manufactures, imports or supplies any substance for use at work to: (a) ensure as far as is reasonably practicable that the substance is safe and without risk to health when properly used; (b) take such steps as are necessary to secure that adequate information is available concerning the use of the substance, to ensure safety and to minimize risk to health when properly used. Future legislation under the Control of Substances Hazardous to Health (COSHH) regulations will introduce statutory duties on employers to implement occupational hygiene measures appropriate to the type of hazard and the level of risk. Manufacturers of hazardous substances have two basic devices to impart the necessary information: (i) the safety data information sheet or bulletin, and (ii) the labels that they put on containers. The European Communities Classification Packaging and Labelling Regulations apply to all substances which are hazardous to health and they stipulate that risk phrases as well as safety phrases have to appear on the labels. All of these regulations refer to the appropriate toxicology testing guidelines which detail *in vivo* studies, using laboratory animals or the experience and knowledge of effects in man, as the criteria by which to classify a chemical's toxicity.

In vitro studies are being introduced into more and more areas of toxicology to assist in such assessments. They will lead to the refinement of procedures using animals and to some reduction of the number of laboratory animals exposed to toxic substances. The total replacement of *in vivo* studies for the assessment of acute systemic, cutaneous and ocular toxicity by *in vitro* procedures does not seem possible at this time. There is no doubt that industry will continue to develop and use *in vitro* techniques to reduce the use of laboratory animals wherever possible.

Acknowledgements

The authors wish to thank the many colleagues who have helped to develop *in vitro* techniques in the Central Toxicology Laboratory, Alderley Park, Macclesfield. We are especially appreciative of Mrs J. Murphy for typing the manuscript and Mr J. Reece for preparing the figures.

References

Benson, V., Clausen, J., Ekwall, B., Hesnten-Pettersen, A., Holme, J., Hogberg, J., Niemi, M. and Walum, E. (1986). *Trends in Scandinavian Cell Toxicology (ATLA)*, **13**, 162–79.

Botham, P. A., Rattray, N. J., Walsh, S. T. and Riley, E. J. (1987). The control of the immune response to contact sensitivity chemicals to cutaneous antigen presenting cells. *Br. J. Dermatol.*, **117**, 1–9.

Crane, R. K. and Wilson, T. H. (1958). *In vitro* method for the study of the rate of intestinal absorption of sugars. *J. Appl. Physiol.*, **12**, 145–6.

Doluisio, J. T., Billups, N. F., Dittert, L. W., Sugitta, E. T. and Swintosky, J. V. (1969). Drug absorption. I. An *in situ* rat gut technique yielding realistic absorption rates. *J. Pharm. Sci.*, **58**, 1196–202.

Duffy, P. A. Flint, O. P., Orton, T. C. and Fursey, M. J. (1986). Initial validation of an *in vitro* test for predicting skin irritancy. *Fd Chem. Toxicol.*, **24**, 517–18.

Dugard, P. H. (1986). Absorption through the skin: Theory, *in vitro* techniques and their application. *Fd Chem. Toxicol.*, **24**, 749–53.

Dugard, P. H., Scott, R. C., Ramsey, J. D., Mawdsley, S. J. and Rhodes, C. (1983). Percutaneous absorption of phthalate esters: *In vitro* experiments on human and rat epidermal membranes. *Human Toxicol.*, **3**, 561.

ECETOX Monograph No. 6 (1985). *Acute Toxicity Testing LD_{50} (LC_{50}), Determination and Alternatives.* Brussels, Belgium, European Chemical Industry Ecology and Toxicology Centre.

Fry, J. R., (1983). A review of the value of isolated hepatocyte systems in xenobiotic metabolism and toxicity studies. In *Animals in Scientific Research: An Effective Substitute for Man?* Ed. Turner, P., pp. 81–90. London, Macmillan.

Gillette, J. R. (1986). On the role of pharmacokinetics in integrating results from *in vivo* and *in vitro* studies. *Fd Chem. Toxicol.*, **24**, 711–20.

Holmstedt, B. (1976). Activities of the European Medical Research Councils with regard to training and research toxicology. *Proc. Eur. Soc. Toxicol.*, **17**, 131.

Kimber, I., Mitchell, J. A., and Giffin, A. C. (1986). Development of a murine local lymph node assay for the determination of sensitising potential. *Fd Chem. Toxicol.*, **24**, 585–6.

King, L. J., Wiebel, F. J. and Zucco, F. (Eds) (1986). Application of tissue culture in toxicology (Proceedings of the Third International Workshop, Urbino, Italy). *Xenobiotica*, **15**, (Nos 8/9).

Lawrence, R. S., Groom, M. H., Ackroyd, D. M. and Parish, W. E. (1986). The chorioallantoic membrane in irritation testing. *Fd Chem. Toxicol.*, **24**, 497–502.

Lhugenot, J. C., Mitchell, A. M., Milner, G., Lock, E. A. and Elcombe, C. R. (1985). The metabolism of DEHP and MEHP in rats: *in vivo* and *in vitro* dose and time dependency of metabolism. *Tox. Appl. Pharm.*, **80**, 11–22.

Oliver, G. J. A., Pemberton, M. A. and Rhodes, C. (1986). An *in vitro* skin corrosivity test—modifications and validation. *Fd Chem. Toxicol.*, **24**, 507–12.

Oliver, G. J. A., Pemberton, M. A. and Rhodes, C. (1988). An *in vitro* model for identifying skin corrosive chemicals. I. Initial validation Toxicol. *In vitro*, **2**(1), 7–17.

Parish, W. E. (1985). Ability of *in vitro* (corneal injury—eye organ—and chorioallantoic membrane) tests to represent histopathological features of acute inflammation. *Fd Chem. Toxicol.*, **24**, 215–27.

Parish, W. E. (1986). Evaluation of *in vitro* predictive tests for irritation and allergic sensitisation. *Fd Chem. Toxicol.*, **24**, 481–94.

Pemberton, M. A., and Oliver, G. J. A. (1986). The measurement of skin irritation *in vitro* using *ex vivo* rabbit skin. *Br. J. Dermatol.*, **115**, 45–6.

Perrier, D. and Gibaldi, M. (1973). Calculation of absorption rate constant for drugs with incomplete availability. *J. Pharm. Sci.*, **62**, 1486–90.

Price, J. B., Parry, M. P. and Andrews, I. J. (1986). The use of the chick chorioallantoic membrane to predict eye irritants. *Fd Chem. Toxicol.*, **24**, 503–505.

Purchase, I. F. H., Koning, D. M., Balls, M., Rhodes, C., Davies, D. N. and Bridges, J. M. (1986). Food chemical toxicology. In *Proceedings of International Conference on Practical in vitro Toxicology*. Reading Univ., pp. 447–517.

Reinhardt, Ch. A., Pelli, D. A. and Zbinden, G. (1985). Interpretation of cell toxicity data for the estimation of potential irritation. *Fd Chem. Toxicol.*, 247–52.

Rhodes, C., Purchase, I. F. H., Pemberton, M. A. and Oliver, G. J. A. (1987). Balanced approach to the assessment of toxicity. *J. Toxicol.*, in press.

Rozmann, K. and Hanninen, O. (Eds) 1986. *Gastrointestinal Toxicology*. Amsterdam, New York and Oxford, Elsevier.

Scott, R. C., Ramsey, J. D., Ward, R. J., Thompson, M. A., Noden, C. and Rhodes, C. (1986a). Percutaneous absorption: *in vitro* assessment. *Fd Chem. Toxicol.*, **24**, 763–4.

Scott, R. C., Thompson, M. A., Ward, R. J., Ramsey, J. and Rhodes, C. (1986b). *In vitro* absorption of l-chloro-2,4-dinitrobenzene (DNCB) through human, hooded rat and mouse epidermis. *Br. J. Dermatol.*, **115**, 47–8.

Spiro, H. (1985). Winning with Archimedian principles. *ATLA*, **13**, 117–22.

Wilson, T. H. and Wiseman, G. (1954). The use of sacs of everted small intestine for the study of the transference of substances from the mucosal to the serosal surface. *J. Physiol.*, **123**, 116–25.

CHAPTER 9

Laboratory methods for assessing carcinogen exposure in man

R. C. Garner

Introduction

The finding, from epidemiological studies, that cancer incidence varies from country to country, from area to area and even from town to town, implies that much human cancer is influenced by environmental factors. It is not the purpose of this chapter to review the literature relating to human epidemiology, this has been done adequately elsewhere (Doll 1978). It is the intention to examine current methodology that could be used to monitor exposure of individuals, possibly at a cellular level, to carcinogens, i.e. those substances responsible for initiating cancer.

The process of cancer induction is complex and certainly not fully understood. Agents known to cause human cancer have been separated into six groups:

1. Irradiation, e.g. X-rays or ultra-violet light.
2. Inorganic metals, e.g. arsenic and nickel.
3. Hormones, e.g. diethylstilbestrol and anabolic steroids.
4. Viruses, e.g. Epstein–Barr virus, herpes virus and human immuno deficiency type.
5. Organic chemicals, e.g. benzidine and vinyl chloride.
6. Physical carcinogens, e.g. asbestos.

Group 5 is probably the largest group in terms of total cancer incidence. Some coming together of groups 4 and 5 is also occurring through the elucidation of the mechanism of action of oncogenes. Of the six classes of carcinogen it could be argued that four exert their effects directly through interaction with DNA, i.e. 1, 2, 4 and 5; the other two groups may have a direct DNA damaging effect but this has yet to be elucidated.

This review will be directed towards examining human exposure to organic chemicals and how this may be monitored using a variety of novel techniques.

Mechanism of carcinogen action

In order to attempt to devise a strategy for monitoring human carcinogen exposure, it is necessary to have some understanding of carcinogen metabolism. Figure 9.1 sets out a commonly accepted scheme for the process. At each point a variety of factors, including nutritional status, species, sex, age and disease status can affect the rate at which each step occurs.

Figure 9.2 shows a more specific example of the complex processes whereby a carcinogen is metabolized in the body. Aflatoxin B_1 (AFB_1) is used as an example because many of the metabolic routes for this mycotoxin have been elucidated. Inter-individual differences in carcinogen susceptibility could be related to the balance between activation to AFB_1-8, 9-oxide and detoxification to metabolites such as AFP_1.

The activation and detoxification of carcinogens occurs in a variety of organs but is probably most extensive in the liver. The enzymes concerned are primarily mono-oxygenases, which catalyze phase 1 metabolism (i.e. oxidation) and phase 2 metabolism (i.e. conjugation) (Gibson and Skett, 1986). Many organic chemical carcinogens appear to be metabolized through similar processes to those described for AFB_1. This process of activation to an electrophilic metabolite has been demonstrated for such diverse structural groups as the nitrosamines, the polycyclic aromatic hydrocarbons, aromatic amines,

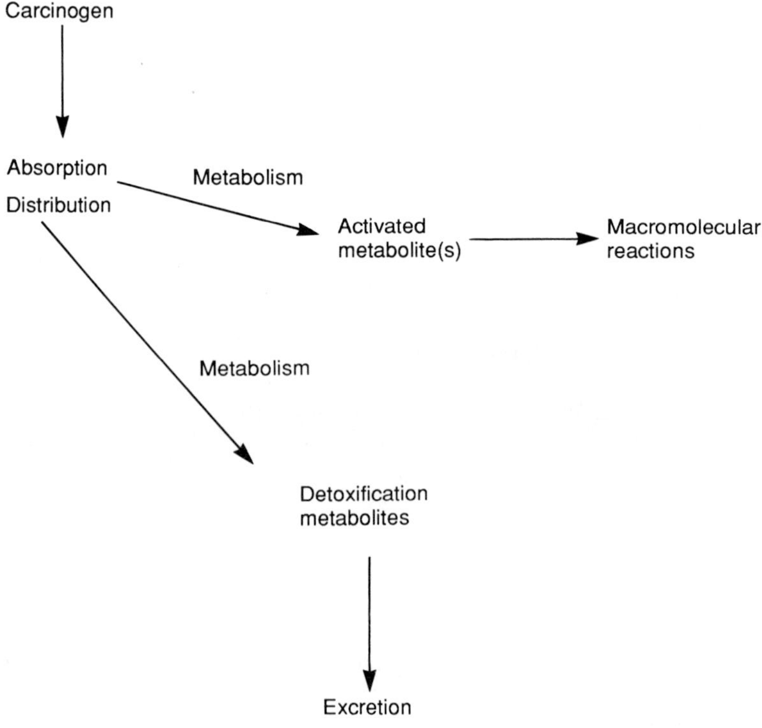

Fig. 9.1 Simplified scheme for carcinogen metabolism

halogenated hydrocarbons, etc. (Miller, 1978). Many macromolecular adducts of these carcinogens have been structurally identified as having some of the mechanisms by which cells recover from such damage.

The interaction of a carcinogen with DNA, a fundamental property of many carcinogens which are activated to electrophiles, is thought to be an important, if not essential, step in the cancer process. Current hypotheses suggest that carcinogen reaction leads to mutational events within the cells, finally leading to a loss of growth control. Molecular biological methods using restriction mapping, base-sequencing and gene

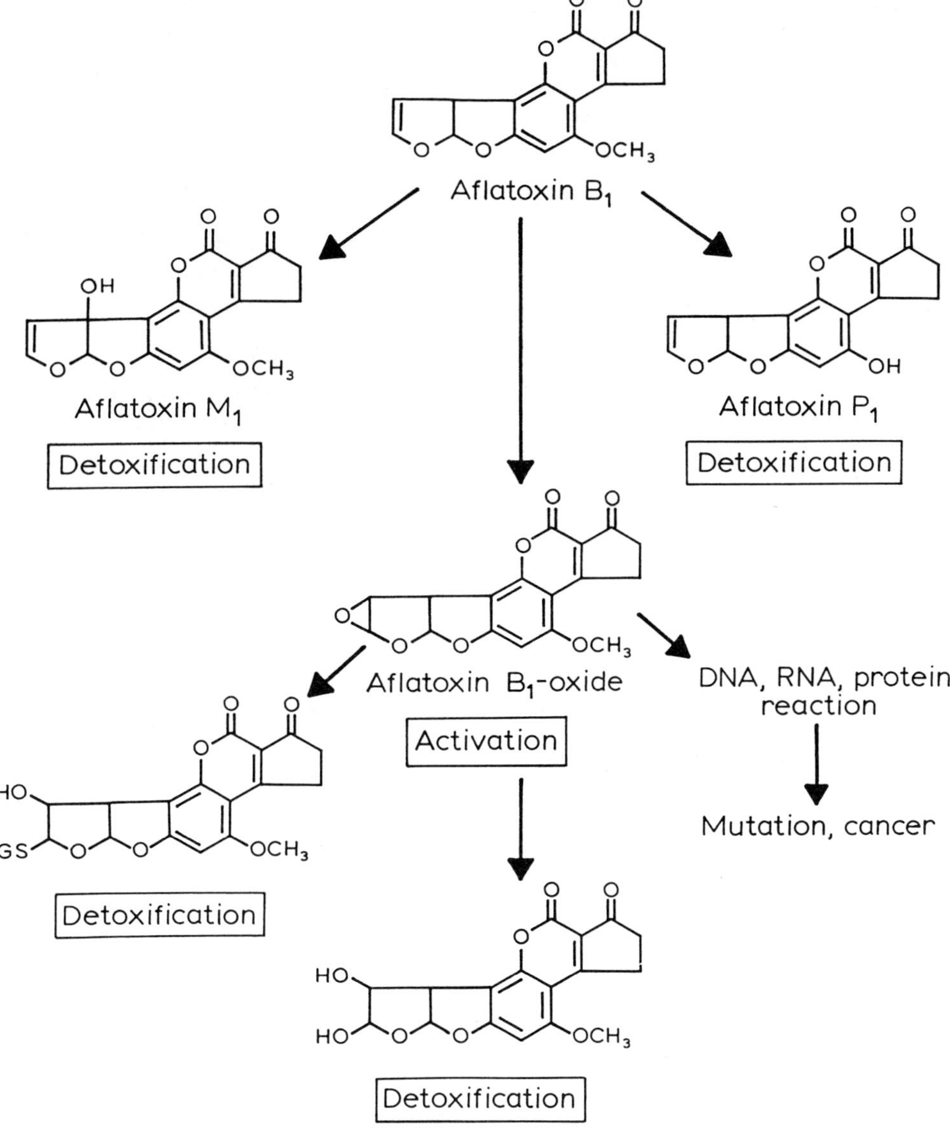

Fig. 9.2 Metabolic activation and detoxification pathways for aflatoxin B_1

probing all support this view of cancer initiation. DNA reaction, therefore, should provide a means of studying critical events for the cancer process at a molecular level in both animals and man.

Why study man and how can this be done?

There is considerable scientific controversy relating to results of animal carcinogenicity experiments and their relevance to man. Examples include the hepatocarcinogenicity of halogenated hydrocarbons (Farber, 1984) and a variety of liver enzyme-inducing agents (Rao and Reddy, 1987). Do these animal results relate to man and human exposure? Until recently it was not possible to contemplate laboratory-based methods that might answer this question. Many of the human risk assessment procedures involve extrapolation from animal data, despite the fact that known differences in pharmacokinetics and pharmacodynamics exist. How sure are we that an animal's response to a high dose of chemical extrapolates back to a low dose? Can we be sure that rates of metabolism, routes of metabolism and macromolecular binding are linearly related to dose? Such questions cannot be answered on man using the same techniques as used routinely in animal experiments. One could not conceive of injecting large doses of radiolabelled compound into people and removing tissues or organs to measure macromolecular binding levels.

This review is, therefore, concerned with non-invasive, rapid, cheap, sensitive, specific and reliable methods for performing studies in man. Several conferences and reviews have been devoted to this topic and the reader should refer to these for an overview of the subject (Garner, 1985). All the methods are designed to move experimental cancer research from an animal-based to a human-based subject. Weinstein has coined the term 'molecular cancer epidemiology' to describe this research area.

MACROMOLECULAR BINDING

Many organic chemical carcinogens have been found to bind to cellular macromolecules after administration to animals. These macromolecular bound species are usually formed at nucleophilic centres as a result of electrophile activation of the carcinogen as outlined above. Numerous studies have been performed demonstrating the binding of carcinogens to DNA and the subsequent repair of such damage. Since reaction of an electrophile can occur with many nucleophilic centres, not only can one demonstrate DNA binding of carcinogens, but binding to RNA and protein also. These latter two have not been as extensively studied because DNA reactions are thought to be of primary importance in initiating tumours. Nevertheless, there are undoubtedly relationships between protein, RNA and DNA binding. Levels of binding to one macromolecule by a carcinogen might be used to develop human monitoring programmes (Perera, 1987b).

Carcinogen binding to macromolecules often occurs to the greatest extent in organs which subsequently develop tumours. However, there is not a one hundred percent relationship between DNA binding and subsequent tumour production. A good example is the case of ethyl-nitrosourea which, although it binds overall to a greater extent to liver DNA than to brain DNA, induces tumours of the brain and not the liver (Preussmann and Stewart, 1984). In this case, it appears that the site of reaction and repair of a particular base, O^6-ethyl-guanine, is important, rather than the mere fact that a DNA binding reaction takes place (Martin and Garner, 1987). In other words, studies concerned with monitoring for carcinogen exposure should recognize that not all DNA reactions appear to be deleterious.

Nevertheless, for certain compounds, there has been shown to be a good correspondence between the level of DNA binding and subsequent tumour induction. In the case of the ionizing radiation, this relationship has led to the recommendation of international safety standards, in which the levels of radiation thought to be biologically acceptable are laid down. In the case of chemical carcinogens, one is striving to develop similar safety standards based on scientifically acceptable criteria. This has so far been difficult to achieve, since there are a plethora of factors which can affect levels of DNA binding as a result of carcinogen exposure. In the case of radiation, the DNA damage induced is direct, i.e. so many rads will induce a certain amount of DNA damage. With chemicals, although one can determine exposure levels, little information exists on what exposure leads to an unacceptable level of DNA damage.

MEASUREMENT OF MACROMOLECULAR BINDING

In animals, the determination of macromolecular binding is by radioactivity measurement after administration of radiolabelled chemical. Provided one can obtain the chemical under investigation in a pure radioactive stable form, then assaying of binding levels is reasonably routine (Martin and Garner, 1987). For human tissues one can only use indirect methods with radiolabelled chemical. Studies have been conducted on human cultured cells or organ explants to demonstrate that human tissue behaves similarly to animal tissue (Gabrielson and Harris, 1985). If DNA adducts of similar structure are obtained, then *a priori* man has the same activating enzymes as the susceptible animal species and is, therefore, susceptible. This can lead to studies which demonstrate AFB_1 can bind to human lung DNA, even though there is no convincing evidence that this mycotoxin is a human lung carcinogen.

Recently, a number of procedures has been developed which could prove useful in the measurement of macromolecular carcinogen bound species in man. These are described below.

Antibody methods

Antibodies can be exquisitely sensitive and specific for their antigen. This fact has been used for many years in clinical chemistry to measure low concentrations of analyte in human body fluids. Antibodies have been raised in sheep, rabbits or goats routinely for clinical immunoassay. The production of antibodies against DNA adducts was first reported in the mid-60s but little attention was paid to these early reports. It has only been with the introduction of monoclonal antibody technology that interest in antibody techniques in general has been rekindled. The early reports concerned the production of antibodies to DNA photoproducts induced by ultra-violet light. DNA is a notoriously poor antigen and immunization procedures had to be adapted to augment the antibody response. The original publications concerned the generation of polyclonal antibodies in rabbits. Little attention appears to have been paid to the sensitivity and specificity of these antibodies. It was only the pioneering work of Erlanger and his colleagues, as well as Poirier, that indicated the vast potential of antibodies for DNA adduct screening (reviewed by Garner, 1985).

A typical scenario in relation to human monitoring would be to obtain antibodies against the adduct of interest, develop an appropriate immunoassay procedure and then analyse for adducts in suitable human tissues or body fluids. Whilst one can spell out the procedures in a few words, the reality is that it may take months, if not years, to carry out the above process. Antibodies can be generated against DNA adducts when bound to DNA, to the isolated DNA adduct or to the carcinogen moiety. Each of these procedures has been used to prepare useful antibodies.

Antibodies against DNA adducts in DNA As already mentioned, immunization with DNA does not generally lead to high antibody titres. The situation can be improved by co-immunizing not only with Freund's adjuvant, but also with methylated bovine serum albumin or methylated keyhole limpet haemocyanin. The co-injection of the protein appears generally to activate the B-cells in the spleen, rendering them more responsive to the weak DNA immunogen. Methylation of the protein enables the DNA to attach electrostatically and so stabilize the complex.

Polyclonal antibodies have been generated utilizing these methods against benzo(a)pyrene-diol-epoxide-reacted DNA (Poirier *et al.*, 1980), acetoxyacetyl-aminofluorene-reacted DNA (Poirier *et al.*, 1977), *cis*-diamminedichloroplatinum reacted DNA (*cis*-DDP-DNA) (Malfoy *et al.*, 1981), DNA-containing thymine dimers generated by ultra-violet light (Strickland, 1985) and aminofluorene-reacted DNA (Kriek *et al.*, 1984). Some of these antibodies have been used to analyse human DNA. In a series of analyses of human lung DNA recovered from surgical specimens, five of the samples had demonstrable benzo(a)pyrene-DNA adducts. All the positive samples were from cigarette smokers. Analysis was by ELISA after immunoaffinity chromatography (Tierney *et al.*, 1986), which appears to be becoming the method of choice for immunoassay.

In another study using a polyclonal antibody to *cis*-DDP-DNA, out of 130 blood samples taken from 63 individuals no false-positive results were obtained, i.e. only lymphocyte DNA from cisplatin-treated individuals had measurable adducts (Poirier, 1984). An increasing accumulation of *cis*-DDP DNA adduct levels was observed in individuals given repeated doses of the drug. Predictivity of adduct levels in exposed peripheral human lymphocytes *in vitro* and adduct levels *in vivo* has been demonstrated (Fichtinger-Schepman *et al.*, 1987).

Monoclonal antibodies have been prepared from benzo(a)pyrene-diol-epoxide-reacted DNA (Perera, 1987a), aflatoxin B_1-reacted DNa and imidazole-opened aflatoxin B_1 (reviewed by Perera, 1987a).

Antibodies against DNA adducts conjugated to proteins Since DNA is such a poor immunogen, a number of workers have prepared conjugates of the isolated DNA-adduct with a protein such as bovine serum albumin. These conjugates have been used to prepare both poly- and monoclonal antibodies.

Polyclonal antibodies have been raised against O^6-methyldeoxyguanosine, guanosine-(8-yl)-acetylaminofluorene, guanosine-(8-yl)-aminofluorene and O^6-ethyldeoxy-guanosine (Garner, 1985). Many of these antibodies have been used to compare adduct levels in animal tissues as measured by radioactivity determinations and results by immunoassay. On the whole, a good correspondence has been obtained between the two procedures using adducted DNA obtained either from treated cell cultures or from animals dosed with the appropriate carcinogen.

Monoclonal antibodies have been obtained using either immunized mice or rats and subsequent cell fusions against O^6-methyl-, O^6-ethyl- and O^6-butyldeoxyguanosine, O^4-

methylthymine and O^2-methylthymine. Oesophageal DNA from patients in high oesophageal cancer incidence areas in China, have been found to have detectable levels of O^6-methylguanine as measured using immunoassay techniques. Control populations had no detectable O^6-methylguanine levels (Umbenhauer *et al.*, 1985).

Antibodies against carcinogens An indirect method to measure carcinogen adducts with DNA utilizes antibodies generated against the carcinogen moiety. Whilst this procedure does enable adducts of differing structure to be detected its non-specificity is a disadvantage when using body samples. Unless the DNA adducts are extensively purified first, then cross-reaction of the antibody will occur with both RNA and protein contaminants. Non-specificity of reaction might be advantageous in 'fishing expeditions' in which one is wanting to find out if a particular organ has any carcinogen adducts contained within it. Antibodies against carcinogens can be either polyclonal or monoclonal and are produced as a result of immunizing with a carcinogen–protein conjugate.

We have extensively studied both animal and human samples to determine if they contain aflatoxin moieties subsequent to aflatoxin exposure. For these studies we have utilized a polyclonal preparation prepared in rabbits, as a result of repeated immunization with an aflatoxin B_1-bovine serum albumin conjugate. Using an ELISA method developed in our laboratories, we have examined human urine obtained from patients in The Gambia. Large variations in antibody inhibitory material were found, indicating that some of the samples did indeed appear to contain aflatoxin or its metabolites (Martin *et al.*, 1984). Whilst it is possible to obtain qualitative measurements using these methods, it has proved impossible to obtain quantitative information on unextracted urine. The reason for this is our finding that different aflatoxin metabolites or DNA adducts have different affinities for the antibody. Antibody binding to these metabolites will reflect not only their concentration but different metabolite affinities. The only way that this particular problem might be resolved would be to purify individual metabolites by chromatography and to assay fractions. This procedure would negate the advantages of immunoassay in terms of speed and simplicity.

Nevertheless, we have used an anti-aflatoxin B_1 monoclonal antibody to make comparisons between macromolecular binding in the rat as assayed by radioactivity and by ELISA (Garner *et al.*, 1988a). The results indicate that it should be possible to measure plasma protein binding in man and extrapolate to estimate levels of DNA binding. Since the antibody preparation was raised against an aflatoxin B_1-protein conjugate, it has proved sensitive to aflatoxin B_1 moieties when bound to protein.

An extension of this approach is that reported by Sun and his colleagues. They have prepared a monoclonal antibody against aflatoxin M_1 and used this to make an immunoaffinity column. When urine from people thought to have been exposed to aflatoxin was passed through the column and the bound material subsequently eluted, aflatoxin M_1 was detected. There was an approximate 1% conversion rate of aflatoxin B_1 to aflatoxin M_1 when measurements were made of aflatoxin B_1 levels in infested beer and the amount of aflatoxin M_1 excreted in the urine (Harris and Sun, 1984). A similar immunoaffinity column procedure has been reported by Groopman and co-workers using an IgM monoclonal antibody to aflatoxin B_1 (Groopman *et al.*, 1985).

Antibody methods—the future

In order for cancer to develop in man there is a requirement, in the majority of cases, for continuous exposure to the carcinogen. Cigarette smoking provides a good example since smoking a single cigarette is not likely to give lung cancer. On the other hand, smoking twenty cigarettes a day for many years increases the risk of contracting cancer of the lung many-fold. Our ability to measure exposure over a long time period is limited by our present methodology. It would be desirable to find some biological disturbance that could reflect the cumulative effect of carcinogens. Our present methods suffer from their inability to provide anything more than single time point estimations of exposure. This is likely to be of little use in determining the reason for individual differences in susceptibility. An approach to try to overcome this problem might be to collect all urine voided by an individual over the period of a month. This urine would be passed through an immunoaffinity column and any of the antigen under study would be trapped. After a month's collection and immunoaffinity chromatography, any bound antigen would be eluted and estimated by immunoassay or physico-chemical procedures. If the immunoaffinity column was prepared from an antibody against a carcinogen–DNA adduct, then it would be possible to measure the amount of this adduct excreted over the period of a month. The adduct amount should reflect the level of DNA damage over the month as well as the rate of repair of this damage. Such measurement could provide an estimation of the total body burden to a carcinogen.

Other estimations of carcinogen body burden could be derived from carcinogen reactions with haemoglobin or other plasma proteins or lymphocyte DNA. Such estimations, which could be by either physico-chemical and/or antibody procedures, could provide the scientist with the necessary information to estimate risk. The measurement of these adducts will depend not only on whether or not they are formed, but also on their stability and longevity. Little attention has to date been obtained on the use of antibody methods to measure carcinogen–protein adducts, but these might prove a more useful method than techniques such as mass spectrometry.

^{32}P-POST-LABELLING ANALYSIS OF DNA

Many carcinogens when they react with DNA give rise to adduct–base conjugates, which have different chromatographic properties when compared with their parent base. This principle has provided a novel procedure which, to date, has been primarily used for adduct analysis in animals. The method, developed by Randerath and his co-workers (Reddy and Randerath, 1986), is to digest enzymically extracted purified-organ DNA down to the monophosphates using DNase-1 and snake venom phosphodiesterase. The total DNA digest is then treated with polynucleotide kinase and [^{32}P]-ATP. The resulting bisphosphates will be labelled with [^{32}P] at the 5′ end. The labelled digest is then subjected to two-dimensional, thin-layer chromatography using PEI-cellulose. Any carcinogen-base biphosphates are likely to have different R_f values compared with the parent bases. Visualization is by autoradiography. A variety of carcinogens has been examined using this technique, including polycyclic aromatic hydrocarbons, benzidine, 2-acetylaminofluorene, safrole and related compounds. Disadvantages of the procedure are that an authentic carcinogen-base monophosphate is required as a chromatographic reference and that many carcinogen-base bisphosphates do not have a sufficiently different R_f value compared with the unreacted bases. This applies particularly to adducts which are substituted with small residues such as methyl and ethyl.

Digestion of lymphocyte DNA from a smoker might give rise to a whole host of spots on TLC, which do not correspond with the normal bases. However, unless the authentic standards are available it is impossible to identify these. The limit of sensitivity of this method is approximately 1 in 10^9 bases, which is at the lower end of the sensitivity scale required to measure nucleic acid adducts in real-life situations. On the other hand, sensitivity is dependent solely on the specific radioactivity of the [^{32}P] used. The method has been used to examine placenta of smokers (Everson *et al.*, 1986) as well as human lung tissue (Garner *et al.*, 1986).

MACROMOLECULAR BINDING MEASURED BY PHYSICO-CHEMICAL METHODS

There have been several published papers on the use of analytical methods to measure macromolecular adducts (Farmer *et al.*, 1986; Bryant *et al.* 1987). Most attention has been directed towards studies on alkylated or arylated haemoglobin using mass spectrometry as the detection method. In this procedure, haemoglobin is obtained from animals or persons exposed to carcinogens, or presumptive carcinogens, and hydrolyzed down to amino acids. Crude separation of the amino acids is carried out by chromatography and the resultant products analysed by mass spectrometry. Many carcinogens have been shown to react with haemoglobin in animal studies, suggesting that this approach may have common application. Since there are nucleophilic sites in haemoglobin these are susceptible to attack by electrophilic carcinogens. As a result, substituted amino acids are obtained which can be quantitated by mass spectrometry (Bryant *et al.*, 1987).

Farmer *et al.* (1986) have reported on studies with ethylene oxide, a direct-acting mutagen. Ethylene oxide reacts with histidine in haemoglobin to yield N-3-(2-hydroxyethyl)histidine. Levels of this substance have been correlated in animal studies with levels of alkylation of liver and testicular DNA. A correspondence has been found between liver binding and haemoglobin binding, implying that the latter can be used to monitor the former. Whilst this approach is an elegant one, very few carcinogens to date have been studied. Reaction with haemoglobin will be dependent on the activated carcinogen actually reaching the target protein. This is going to be an unlikely event for highly reactive electrophiles. On the other hand, a carcinogen such as 4-aminobiphenyl, and possibly other aromatic amine carcinogens, because they can be activated by red blood cell peroxidases, actually attack the haemoglobin *in situ*. Tannenbaum and his colleagues have shown that they can detect 4-aminobiphenyl adducts with haemoglobin using mass spectrometry in animals not thought to be knowingly exposed to 4-aminobiphenyl (Green *et al.*, 1984).

In addition to mass spectrometry of alkylated haemoglobins, one can use similar techniques to analyse for excreted macromolecular adducts in the urine. The sensitivity of the instrument means that very low levels of adduct might be detected after low-dose carcinogen exposure. 7-Methylguanine has been demonstrated in the urine of animals dosed with either dimethylnitrosamine (Craddock and Magee, 1967) or the precursors aminopyrene and sodium nitrate (Gombar *et al.*, 1983).

A combination of immunoaffinity chromatography for concentration and mass spectrometry for identification might prove a useful approach.

Fluorescence spectroscopy provides a further method for examining adduct levels in human tissues. A number of carcinogens, such as benzo(a)pyrene and aflatoxin B_1, have characteristic and strong fluorescence spectra. With the introduction of powerful computers and Fourier transform analysis to separate signal from noise, adduct levels as low as 1 in 10^6 to 1 in 10^8 can be detected. It has been possible to demonstrate aflatoxin B_1–guanine in the urine of persons thought to be exposed to aflatoxin B_1 in Kenya (Autrup *et al.*, 1983). DNA samples from alveolar macrophages, obtained by lavage, have been found to have detectable levels of benzo(a)pyrene–DNA adducts (Shamsuddin *et al.*, 1985). Not only do certain carcinogens have characteristic fluorescence spectra, but so do some base adducts. O^6-methylguanine, for example, has a stronger fluorescence spectrum than guanine. The ethano–adenine adduct formed on reaction of vinyl chloride epoxide with adenine also has a characteristic spectrum. Such physico-chemical properties can be exploited provided these are known in advance.

HUMAN TISSUES THAT CAN BE SAMPLED NON-INVASIVELY

Resourcefulness is needed to find tissues that can be used for monitoring purposes. Urine and faeces provide the easiest, if not the most pleasant, of body fluids to obtain. Most subjects do not object to providing specimens for analysis. Difficulties come in the storage of large numbers of samples or in the volume of samples for population-based studies. Blood samples are a little more difficult to obtain, requiring trained personnel, but are often removed for other purposes, such as clinical chemistry. Other cells that may be used for analysis include those from the buccal lining of the mouth and exfoliated bladder cells, which have sloughed off into the urine. These latter can be obtained by centrifugation of the voided urine sample.

From all the cell types obtained as above, it is possible, using micro-methods, to isolate sufficient quantities of DNA, RNA and protein for analysis. Limits of sensitivity of detection procedures need, however, to be in the nanogram to picogram range.

Summary

Various techniques have been described which enable one to estimate levels of carcinogen–macromolecular adducts in human tissues. Most of these procedures have sufficient sensitivity and specificity to estimate exposure in real, as opposed to artificial, situations. Very few of the methods have, as yet, been used systematically and on a large scale to perform population-based studies. More effort and resources need to be applied to this research area in the future if we, as scientists, are going to be able to relate animal-based studies to human risk. In addition, the methods described should enable us to assess the importance of environmental chemicals in the aetiology of human cancer.

References

Autrup, H., Bradley, K. A., Shamsuddin, A. K. M., Wakhisi, J. and Wasunna, A. (1983). Detection of putative adduct with fluorescence characteristics identical to 2,3-dihydro-2-(7′-guanyl)-3-hydroxy aflatoxin B_1 in human urine detected in Murang'a district, Kenya. *Carcinogenesis*, **4**, 1193–5.

Bryant, M. S., Skipper, P. L., Tannenbaum, S. R. and Maclure, M. (1987). Haemoglobin adducts of 4-aminobiphenyl in smokers and non-smokers. *Cancer Res.*, **47**, 602–8.

Craddock, V. M. and Magee, P. N. (1967). Effect of administration of the carcinogen dimethylnitrosamine on urinary 7-methylguanine. *Biochem. J.*, **104**, 435–40.

Doll, R. (1978). An epidemiological perspective of the biology of cancer. *Cancer Res.*, **38**, 3573–83.

Everson, R. B., Randerath, E., Santella, R. M., Cefalo, R. C., Avitts, T. A. and Randerath, K. (1986). Detection of smoking-related covalent DNA adducts in human placenta. *Science*, **231**, 54–7.

Farber, E. (1984). Chemical carcinogenesis: a current biological perspective. *Carcinogenesis*, **5**, 1–5.

Farmer, P. B., Bailey, E., Gorf, S. M., Tornquist, M., Osterman-Golker, S., Kautianen, A. and Lewis-Enright, D. P. (1986). Monitoring human exposure to ethylene oxide by the determination of haemoglobin adducts using gas chromatography–mass spectrometry. *Carcinogenesis*, **7**, 637–40.

Fichtinger-Schepman, A. M. J., van Oosterom, A. T., Lohman, P. F. M. and Berends, F. (1988). Interindividual human variation in cisplatin sensitivity, predictable in an *in vitro* assay. *Mutation Res. Lett.*, in press.

Gabrielson, E. W. and Harris, C. C. (1985). Use of cultured human tissues and cells in carcinogenesis research. *J. Cancer Res. Clin. Oncol.*, **110**, 1–10.

Garner, R. C. (1985). Assessment of carcinogen exposure in man. *Carcinogenesis*, **6**, 1071–8.

Garner, R. C., Dvorackova, I. and Tursi, F. (1988a). Immunoassay procedures to detect exposure to aflatoxin B_1 and benzo(a)pyrene in animals and man at the DNA level. *Int. Arch. Occup. Environ. Health*, **60**, 145–50.

Garner, R. C., Tierney, B. and Phillips, D. H. (1988b). A comparison of ^{32}P-postlabelling and immunological methods to examine human lung DNA for BP adducts. *International Conference on Detection Methods for DNA-damaging Agents in Man: Applications in Cancer Epidemiology and Prevention*, in press.

Gibson, G. G. and Skett, P., (1986) *Introduction to Drug Metabolism*. London, Chapman and Hall.

Green, L. C., Skipper, P. L., Turesky, R. J., Bryant, M. S. and Tannenbaum, S. R. (1984). *In vivo* dosimetry of 4-aminobiphenyl in rats via a cysteine adduct in haemoglobin. *Cancer Res.*, **44**, 4254–9.

Groopman, J. D., Donahue, P. R., Zhu, J., Chen, J., Wogan, G. N. (1985). Aflatoxin metabolism and nucleic acid adducts in urine by affinity chromatography. *Proc. natn. Acad. Sci. USA*, **82**, 6492–6.

Gombar, C. T., Lubroff, J., Strehan, G. D. and Magee, P. N. (1983). Measurement of 7-methylguanine as an estimate of the amount of dimethylnitrosamine formed following administration of aminopyrene and nitrite to rats. *Cancer Res.*, **43**, 5077–80.

Harris, C. C. and Sun, T. (1984). Multifactorial etiology of human liver cancer. *Carcinogenesis*, **5**, 697–701.

Kriek, E., Welling, M. and van der Laken, C. J. (1984). Quantitation of carcinogen–DNA adducts by a standardised high-sensitivity enzyme immunoassay. In *Monitoring Human Exposure to Carcinogenic and Mutagenic Agents*, Eds Berlin, A., Draper, M., Hemminki, K. and Vainio, H., IARC Scientific Publication No. 59, pp. 297–305. Lyon, France, IARC.

Malfoy, B., Hartmann, B., Macquet, J-P. and Leng, M. (1981). Immunochemical studies of DNA modified by *cis*-dichlorodiammineplatinum(II) *in vivo* and *in vitro*. *Cancer Res.*, **41**, 4127–31.

Martin, C. N., Garner, R. C., Tursi, F., Garner, J. V., Whittle, H. C., Ryder, R. W., Sizaret, P. and Montesano, R. (1984). An enzyme-linked immunosorbent procedure for assaying aflatoxin B_1. In *Monitoring Human Exposure to Carcinogenic and Mutagenic Agents*, Eds Berlin, A., Draper, M., Hemminki, K. and Vainio, H. IARC Scientific Publication No. 59. Lyon, France, IARC.

Martin, C. N. and Garner, R. C. (1987). The identification and assessment of covalent binding *in vitro* and *in vivo*. In *Biochemical Toxicology—A Practical Approach*, Eds Snell, K. and Mullock, B., pp. 109–26. Oxford, IRL Press.

Miller, E. C. (1978). Some current perspectives on chemical carcinogenesis in human and experimental animals: Presidential Address. *Cancer Res.*, **38**, 1479–96.

Perera, F. P. (1987a). Molecular cancer epidemiology: a new tool in cancer prevention. *J. Natn. Cancer Inst. USA*, **78**, 887–98.

Perera, F. P. (1987b). The significance of DNA and protein adducts in human biomonitoring studies. *Mutation Res.*, in press.

Poirier, M. C., Yuspa, S. H., Weinstein, I. B. and Blobstein, S. (1977). Detection of carcinogen–DNA adducts by radio-immunoassay. *Nature*, **270**, 186–8.

Poirier, M. C., Santella, R., Weinstein, I. B., Grunberger, D. and Yuspa, S. H. (1980). Quantitation of benzo(a)pyrene–deoxyguanine adducts by radio-immunoassay. *Cancer Res.*, **40**, 412–16.

Poirier, M. C. (1984). The use of carcinogen–DNA adduct antisera for quantitation and localisation of genomic damage in animal models and the human population. *Env. Mutagenesis*, **6**, 879–87.

Preussmann, R. and Stewart, B. W. (1984). N-nitro-carcinogens. In *Chemical Carcinogens*, Ed. Searle, C. E., 2nd edn, Vol. 2, ACS Monograph 182, pp. 643–828. Washington DC, American Chemical Society.

Rao, M. S., and Reddy, J. K. (1987). Peroxisome proliferation and hepatocarcinogenesis. *Carcinogenesis*, **8**, 631–6.

Reddy, M. V. and Randerath, K. (1986). Nuclease P_1-mediated enhancement of sensitivity of ^{32}P-postlabelling test for structurally diverse DNA adducts. *Carcinogenesis*, **7**, 1543–51.

Shamsuddin, A. K. M., Sinopoli, N. T., Hemminki, K., Boesch, R. R. and Harris, C. C. (1985). Detection of benzo(a)pyrene–DNA adducts in human white blood cells. *Cancer Res.*, **45**, 66–8.

Strickland, P. T. (1985). Immunoassay of DNA modified by ultraviolet radiation: a review. *Env. Mutagenesis*, **7**, 599–607.

Tierney, B., Benson, A. and Garner, R. C. (1986). Immunoaffinity chromatography of carcinogen DNA adducts with polyclonal antibodies directed against benzo(a)pyrene-diol-epoxide–DNA. *J. Natn. Cancer Inst. USA*, **77**, 261–7.

Umbenhauer, D., Wild, C. P., Montesano, R., Saffhill, R., Boyle, J. M., Huh, N., Kirstein, U., Thomale, J., Rajewsky, M. F. and Lu, S. H. (1985). O^6-ethyldeoxyguanosine in oesophageal DNA among individuals at high risk of oesophageal cancer. *Int. J. Cancer*, **36**, 661–5.

CHAPTER 10

Use of engineered strains of *Escherichia coli* as an alternative to biological assays for available amino acids in foods and feedstuffs

J. W. Payne

Introduction

There is a pressing need to develop rapid *in vitro* methods for determining protein nutritional value of human foodstuffs and animal feeds. New methods are needed to complement, or at times to replace, the classical animal (mainly rat) assays, which are at best too expensive and protracted, and at worst simply inapplicable to particular analytical problems. Over the years, various chemical and microbiological procedures have been developed as alternatives to animal (biological) assays but none is without its disadvantages. As a background against which to consider the *Escherichia coli* assay we have developed, it is convenient first to describe briefly the principles, strengths and weaknesses of current, biological, chemical and microbiology assays. A more detailed discussion of these alternative procedures has been given elsewhere (Payne and Tuffnell, 1980; Tuffnell, 1983).

Assays for amino acid availability

BIOLOGICAL METHODS

Biological procedures, mainly using animals but occasionally involving humans, do not provide direct quantitation of protein or of individual amino acids; rather they give a measure of these components in terms of their nutritional utilizability. It is, of course, precisely this measure that is most commonly required, and it is important to bear this feature in mind when comparison is made with analytical data obtained from alternative procedures. Thus, animal studies provide an index to the nutritional availability of protein amino acids (Bigwood, 1972; Porter and Rolls, 1973; Friedman, 1975; Bodwell, 1976; Cole *et al.*, 1976; Bodwell *et al.*, 1981; Gruenwadel and Whitaker, 1985).

Discrepancies between the results from biological and other assays may reflect the complement of unavailable amino acid residues resulting from incomplete digestion and/or absorption. Unavailable amino acid residues are commonly found in native protein foods but their content is frequently increased during food processing, particularly when heat treatment is involved. In addition, it is a matter of continuing debate as to the extent to which differences in digestive capacities, patterns of absorption, actual nutritional requirements, response to anti-nutritional factors present in samples, etc. between humans and, for example, rats or dogs, may invalidate the extrapolation of results from animal studies to humans (Porter and Rolls, 1973).

The most commonly used animal procedures are ones involving measurement of growth or of nitrogen balance. The basic principle of the procedures is that if any one of the essential amino acids is deficient in the diet then net protein synthesis and thus growth cannot occur. It is therefore the extent of availability of the first nutritionally-limiting amino acid that governs the overall nutritional value of any protein sample. In this context, possible variations in the relative nutritional value of a given quantity of available amino acid present in either free form, as a mixture of small peptides, or as part of a protein, have been discussed elsewhere (Payne and Tuffnell, 1980).

In one of the most commonly used growth procedures, the protein efficiency ratio (PER) method, a group of rats is given the test sample as the sole protein source comprising about 10% of the diet, for about 10–30 days. In a modified assay, the net protein ratio (NPR) is determined. This includes a measure of the dietary protein component required for maintenance and is estimated from the weight loss suffered by a comparable group of rats fed a protein-free diet. NPR is a better measure of protein quality than PER and finds application as a screening test where differences between protein sources are needed. Arguably the most reliable assay in this class gives the relative protein value (RPV). Here the rate of body weight change of rats fed various levels of test protein or reference protein (commonly casein) is determined.

Nitrogen balance methods provide the most detailed information on protein quality, including measures of digestibility and utilization. They have the advantage of requiring shorter experimental times than weight gain methods but, on the other hand, they are more laborious, requiring measurement of nitrogen intake and faecal and urinary-nitrogen excretion. True digestibility (TD) and biological value (BV) are primary characteristics, whereas net protein utilization (NPU) is a measure derived from the previous two. To determine TD, the actual protein nitrogen absorbed is obtained by subtracting that excreted from that consumed, after allowing for endogenous nitrogen excretion by measurements using a control group fed a protein-free diet. The BV of a protein represents that fraction of digested nitrogen retained in the body and provides a measure of the efficiency of the absorbed protein for growth and maintenance. Thus, BV indicates the quality of the protein actually digested and absorbed, so that a protein may be very poorly digested but potentially of high nutritional value. In practice, such a protein would be of little value nutritionally and, consequently, one of the best indices of true nutritional worth is the net protein utilization (NPU) measure, which makes allowance for both amino acid composition and digestibility.

In summary, many theoretical objections can be levelled against current animal assays, despite their apparent fundamental nature. They are also expensive, long-term procedures that require relatively large quantities of sample. *A priori*, therefore, they are excluded from application in a wide range of situations for which assays of protein nutritional quality and amino acid availability are needed.

CHEMICAL METHODS

Many chemical procedures have been devised as alternatives to biological methods. Commonly, they provide only indirect measurement of protein or amino acids, for example, as used in certain screening procedures. Nevertheless, they can be useful in identifying interesting or anomalous samples, although the results may need to be corroborated by applying more direct, nutritional criteria. Generally, chemical methods are relatively fast, which may allow rapid throughput of samples and possibilities of automation. However, they very commonly lack absolute specificity and may have low sensitivity. A common problem is that the content of any amino acid determined chemically may correlate poorly with the nutritional score for the component. This may arise through lack of specificity of the chemical method, as mentioned. Additionally, acid hydrolysis is often used in chemical methods and this may lead to the release of amino acid residues that would be resistant to enzymic digestion *in vivo* and thus not be absorbed and utilized; this difference is commonly observed in samples that have been processed or heat-treated resulting in chemically modified proteins (Finot *et al.*, 1977; Waller and Feather, 1983). On the other hand, acid hydrolysis can break down certain residues that may lead to chemical underestimates of their true nutritional value. Chemical assays would over-estimate nutritional values of samples of poor digestibility and those in which the presence of anti-nutritional factors would lead to impaired protein absorption and utilization *in vivo*.

As mentioned previously, the nutritional quality of different foodstuffs is often dependent on their content of a specific amino acid, commonly lysine, methionine or tryptophan, or less commonly sulphur amino acids in general, threonine, isoleucine or leucine. Consequently, considerable effort has gone into devising chemical assays for these critical amino acids (see for example Concon, 1975). However, these assays are often rather non-specific, e.g. perhaps the most widely applied assay for lysine, which uses 2,4-dinitrofluorobenzene, is essentially an assay for amino groups (Carpenter, 1960; Erbersdobler and Anderson, 1983).

Overall, therefore, chemical methods may prove to be satisfactory in certain instances, but there are various reasons why they often provide an unreliable predictive measure of nutritional value.

MICROBIOLOGICAL METHODS

Microorganisms are widely used as analytical tools, frequently as the first choice for assaying vitamins and antibiotics, and to a lesser extent for amino acids and proteins (Kavanagh, 1972; Board and Lovelock, 1975; Hewitt, 1977). For the latter substances, they effectively bridge the gap between the biological and chemical methods considered above, providing comparable speed, simplicity, sensitivity and potential for automation to the chemical methods, whilst possessing biological features resembling the former procedures. Microorganisms have been used to measure total acid or enzymic hydrolysates of proteins, their relative nutritive values (RNV), the nitrogen content of samples and amounts of specific amino acids.

Many microbial species have been used including bacteria, fungi and protozoa (Payne and Tuffnell, 1980; Landers and Gierhart, 1985). The test organism usually has a growth requirement for one or more specific nutrients and is used in methods that measure a substrate-dependent growth response.

To date, a number of problems exist with respect to microbial assays for available amino acids. First, is the question of the correlation between the microbial results and those obtained with laboratory animals and man. Extensive comparative data are needed to answer this question and this can be a major undertaking. How meaningful such comparisons can be depends in part upon the reliability in response of the microorganism to the test nutrient, and for commonly-used species this is a recognized problem, for many are nutritionally fastidious and their exact requirements poorly documented and understood. It is certainly true that most of the potential advantages mentioned above for a microbiological assay are not being realized in conventional assays for amino acids because of deficiences in the test organisms (see Payne and Tuffnell, 1980, for a fuller discussion of this topic). Thus, until recently, naturally-occurring strains of microorganisms, adapted to particular environments, with their associated idiosyncratic biochemistry and physiology, have been the only source of test organism, and workers have been forced to fashion assays within these evolutionary-imposed limitations. It resembles the situation and inherent limitations that would apply, if in picking up a flint one were obliged to use it directly as an axe head without the opportunity to shape or sharpen it. However, the advent of molecular genetics and recombinant DNA technology now provides the opportunity, in principle at least, to tailor a test organism to the specific demands of an assay, rather than the other way round. It is this appreciation that has been at the basis of our attempts to develop the novel, *in vitro* assay using *E. coli* described here.

Allied to the molecular genetical approach aimed at producing an improved test organism has been our intent to apply standard biochemical principles aimed at developing an improved type of assay. This feature is best considered by first briefly reviewing the standard types of microbial assay that are used at present.

The two most commonly used methods of microbiological assay are the turbidimetric tube method and the agar diffusion (or plate) method. Both procedures involve quantitative comparison between the growth response of the organism to a standard and to the test sample. Growth is measured in a tube by increase in turbidity, or on the surface of an agar plate from the size of the growth zone around a suitable deposit of the essential nutrient. With coloured or turbid samples, alternative procedures are sometimes used, e.g. titrimetry, in which metabolic production of acid is used as an index of growth, or gravimetric assay of actual cell mass. Practical advice and critical appraisal of these microbiological methods is contained in a number of reviews (Kavanagh, 1972; Bolinder, 1972; Hewitt, 1977; Payne and Tuffnell, 1980; Ford, 1981; Landers and Gierhart, 1985).

Development of *E. coli* Assay

BACKGROUND

As part of a long-term project aimed at characterizing the molecular mechanisms of peptide transport in microorganisms and relating the information to the rational design of novel antibiotics (Payne, 1980, 1986; Matthews and Payne, 1980), we routinely used turbidimetric measurement of the growth response of appropriate amino acid auxotrophs of *E. coli* to defined peptides as an indicator of their transport. At this time, colleagues in the Department of Botany, Durham University, were seeking to measure the availability of certain amino acids in various cereal and legume cultivars, as part of a programme aimed at producing nutritionally improved varieties. Thus, it was that we came to consider the possibility of trying to use certain of our *E. coli* auxotrophs to predict the nutritive value of various plant materials.

TURBIDIMETRIC GROWTH TUBE ASSAYS

When we came to make comparisons between the characteristics of *E. coli* and of those microorganisms conventionally used in assays for amino acid availability, the intrinsic advantages of *E. coli* were apparent.

Firstly, *E. coli* offers practical advantages. It can be grown simply and reproducibly in minimal salts media, in contrast to the more complex media and rigorous culture procedures usually needed with the fastidious assay organisms. *E. coli* has a short division time and the nature of the inoculum is relatively unimportant (Bell *et al.*, 1977). Secondly, *E. coli* has physiological advantages. Thus, it is a normal constituent of the microbial flora of the human gut, with well-characterized transport systems for peptides and amino acids, which in many respects are similar to the absorption systems found in the mammalian intestine (Matthews and Payne, 1980). It therefore has properties that make it a particularly suitable tool to use for assessing nutritional value of proteins. *E. coli* also shows equal growth response to equivalent amino acid residues whether free or peptide-bound (Bell *et al.*, 1977), an essential requirement for a reliable assay and yet a feature known to be variable for many fastidious assay organisms (see Matthews and Payne, 1975, and Payne, 1980, for further discussion of these features).

E. coli lacks extracellular protease and peptidase activity. Consequently, when used as a test organism, samples need first to be subjected to a hydrolysis (digestion) step *in vitro*. This proteolysis must be sufficiently extensive to yield products with a small enough Stokes radius to pass through the porins of the outer membrane of the bacterium, via which they gain access to their specific permeases in the inner, cytoplasmic membrane; the limiting size for uptake corresponds to a typical pentapeptide of molecular weight ca. 650 (Payne and Gilvarg, 1968; Alves *et al.*, 1985).

Various criteria need to be applied during development of an *in vitro* proteolysis step. For reasons mentioned earlier, enzymic hydrolysis is necessary and this should match the outcome of the normal digestion process as closely as possible. The procedure should be cheap, effective and convenient and of general applicability, although it is to be expected that modifications may be needed with some types of sample. To this end, the effectiveness of several enzyme combinations on various substrates was monitored, using pH stat titrations to determine the rate and extent of hydrolysis and gel chromatography to establish the size range of cleavage fragments (Bell *et al.*, 1977; Payne *et al.*, 1977). A procedure involving initial milling and sieving of the samples followed by sequential incubation with pronase and mixed intestinal peptidases has been adopted (Bell *et al.*, 1977; Payne *et al.*, 1977; Tuffnell and Payne, 1985).

TURBIDIMETRIC TUBE ASSAY FOR AVAILABLE LYSINE

Lysine is an essential amino acid that is the first limiting amino acid in many foods, especially cereals. Furthermore, lysine is particularly susceptible to chemical modification with loss of nutritional availability during processing and heat treatment of foods (Waller and Feather, 1983), making it important to monitor the level of available lysine during such treatments.

No organism suitable for assaying lysine occurs naturally. However, we have shown that a lysine auxotroph of *E. coli*, M26–26, can be used successfully to assay lysine in a variety of samples (Bell *et al.*, 1977; Payne *et al.*, 1977). The mutant, blocked in the last step of lysine biosynthesis (diaminopimelic acid decarboxylase) is unable to respond to any biosynthetic precursors of lysine; its growth is, therefore, dependent on the amount of available lysine in the test sample. This provides an absolutely specific assay for lysine. In a typical assay, an inoculum (ca. 5×10^6 organisms) is added to a minimal salts glucose medium (2.0 ml) supplemented with lysine standards or protein digests, etc. and incubated at 37°C for 18 h before measuring absorbance at 660 nm. An essentially linear calibration curve is obtained for the approximate range 0–10 $\mu g\ ml^{-1}$ lysine (Payne *et al.*, 1977). The growth response is identical for a given amount of lysine whether it is free or peptide-bound, and it is also unaffected by the presence of other nutrients in the medium. When the assay was applied to digests of various legume and maize meals, good comparison was found with lysine values determined by automated amino acid analysis; the nutritionally available lysine (determined in the *E. coli* assay) being $>75\%$ of the chemical value for each sample (Payne *et al.*, 1977).

In later studies in other laboratories (Anantharaman *et al.*, 1983a,b), comparison was made between lysine availabilities determined using the *E. coli* M26–26 assay and standard chemical and rat procedures. Samples studied included lyophilized cod fillet, desalted peanut flour and wheat flour, and various skim milk products (spray dried, roller dried and lactose-treated and lyophilized). A very high degree of comparability was found between the *E. coli* and chemical methods and also very good correlation with the rat bioassay. The authors concluded that:

> . . . the *E. coli* method is simple, rapid and simulates the normal utilization of digested protein foods *in vivo*. . . . results demonstrate the potential of *E. coli* M26–26 as a suitable alternative to rat bioassay for determination of lysine availability in food proteins. . . . the microbiological determination of nutritionally available lysine, in foodstuffs that have undergone early Maillard reactions (as in milks), using the lysine-requiring *E. coli* mutant M26–26 is a suitable alternative to chemical determinations.

COLORIMETRIC ENZYME ASSAY FOR LYSINE

In a normal turbidimetric assay, nutrient-dependent cell multiplication is measured. In an improved development, we used strain M26–26 to measure lysine-dependent synthesis of the inducible enzyme β-galactosidase (Tuffnell, 1983; Tuffnell and Payne, 1985). With this approach sensitivity can be increased by ca. 10^2–10^3 and the assay time decreased to about 2 h.

The assay procedure currently used has been modified somewhat from the original (Tuffnell and Payne, 1985) and is as follows. Bacteria are grown to log phase in a minimal salts medium minus citrate with 0.5% glycerol as carbon source (to prevent catabolite repression, see later) plus required amino acids (0.5 mM), and harvested by membrane filtration. Washed cells are resuspended in the same medium (minus lysine) and incubated at 37°C for 60 min to deplete the endogenous lysine pool. Lysine-starved bacteria are diluted to 10^8 cells ml^{-1}, 10 mM isopropyl-β-D-thiogalactopyranoside (IPTG) is added as inducer (to give 0.5 mM, or a saturating level), and incubation continued for 15 min at 37°C. Samples (0.5 ml) of induced bacteria are added to tubes containing 1.5 ml of lysine standard or test sample and the tubes incubated for 90 min at 37°C to permit synthesis of lysine-dependent β-galactosidase. Bacteria are cooled to 25°C in a water bath for 5 min and then permeabilized by adding PK disruption medium (40 μl) and vortexing the mixture for 5 sec. After 10 min, 0.2 ml 10 mM *o*-nitrophenyl-β-D-galactopyranoside is added as substrate, the tubes are incubated for 15 min and the enzymic reaction stopped by addition of 0.3 ml 2 M sodium carbonate. The tubes are vortexed hard to clear the suspension and left 30 min to achieve a stable absorbance. The importance of this final step is indicated by the results shown in Fig. 10.1 for two representative concentrations of lysine.

A typical calibration curve for lysine using the above procedure is shown in Fig. 10.2. A linear response between absorbance and amount of lysine is found for the range 0.5–3 nmol lysine. With this degree of sensitivity, about 0.1 mg of sample containing 50% protein would be sufficient to assay for lysine. This makes it particularly attractive for use in screening programmes for new plant varieties where limited sample material is available. In this test system the calibration curve does not pass through the origin. A similar observation is made when different amino acid auxotrophs (e.g. Met and Trp) are used in the same assay (G. M. Payne, J. W. Payne and J. M. Tuffnell, unpublished results). Although several molecular processes could be imagined as contributing to this effect, we prefer not to speculate on possible causes.

In developing this assay, various aspects of the procedure were explored as potential sources of error (Tuffnell, 1983; Tuffnell and Payne, 1985). One difficulty can arise from the fact that the *lac* operon (which includes the gene for β-galactosidase) is subject to catabolite repression. This mechanism is mediated through the pool level of cAMP, which in turn is controlled by the type of carbon source. In consequence, it was shown earlier (Tuffnell and Payne, 1985) that if glucose is present in the test sample it can lead to decreased synthesis of β-galactosidase. This feature is illustrated in Fig. 10.3. Similar

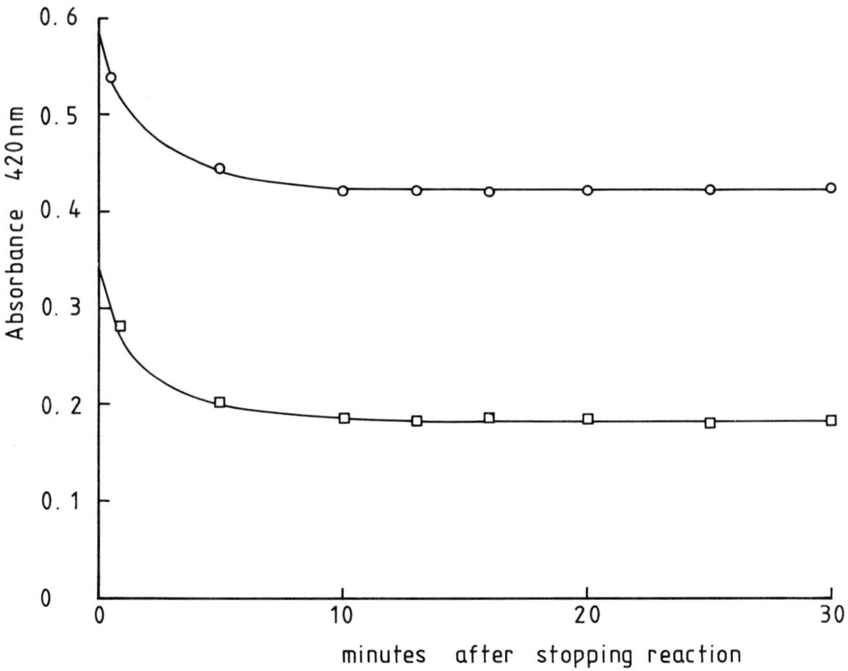

Fig. 10.1 Stabilization of absorbance in β-galactosidase assay. After addition of Na_2CO_3 to stop the enzymic reaction, samples were vortexed and the absorbances of the sample tubes were measured periodically. Samples contained, ○——○, 2 nmol lysine, and □——□, 1 nmol lysine

results are obtained when different amino acid auxotrophs (e.g. Met and Trp) are used (G. M. Payne and J. W. Payne, unpublished results). Thus, if there is reason to suspect that glucose (or other potential catabolite repressor molecules) is present to a significant extent in samples, then this should be tested for by checking the additivity of response of given amounts of test sample and free lysine (or other test amino acid). If necessary, glucose can be removed by a brief pre-incubation with glucose oxidase just prior to assay. (Typically, commercial glucose oxidase is available at 1000 units ml^{-1} of which 1 unit will oxidise 1 μmol glucose min^{-1} at 35°C. Thus, if glucose is present at a 10-fold molar excess over the test amino acid a pre-incubation of 1 min with 0.5 μl enzyme will remove the sugar.) An alternative solution to this problem is discussed below.

In this assay, it was shown that the naturally-occurring lysine analogue δ-hydroxylysine can not substitute for lysine (Tuffnell and Payne, 1985). Furthermore, in agreement with the findings of Anantharaman and colleagues mentioned above, it was shown for the enzyme assay also (Tuffnell and Payne, 1985) that M26–26 effectively gave a measure of available lysine and failed to utilize heat-damaged, Maillard-type products. In this regard, therefore, *E. coli* responds in a manner similar to the young rat, which is used routinely to assay available lysine.

The method was used to assay the lysine content of (a) several pure proteins

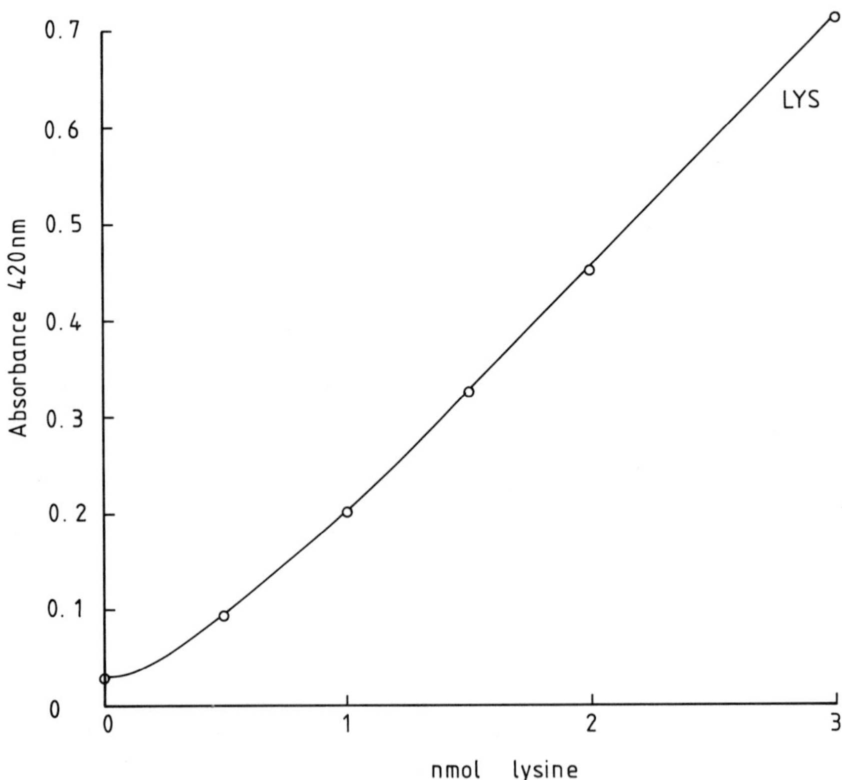

Fig. 10.2 Lysine calibration curve. *Escherichia coli* M26–26 was incubated with the indicated amounts of lysine and β-galactosidase activity was measured as described in the text

(chymotrypsinogen and myoglobin), (b) a range of high-protein feed meals (ground nut meal, soya bean meal and fish meal) and (c) various rice cultivars (Tuffnell and Payne, 1985). Good agreement was obtained between the *E. coli* results, and (a) the theoretical lysine values based on protein amino acid sequence data, (b) the biological results from rat or chick assays obtained in a series of collaborative trials on the feed meals, and (c) total lysine values for the rice samples.

NEW AUXOTROPHIC STRAINS OF *E. COLI*

Using strain M26–26 as parent, we early on selected for methionine and tryptophan mutants and produced a collection of strains in which these requirements were present either alone or in combination (J. M. Tuffnell and J. W. Payne, unpublished results; Tuffnell, 1983). These were tested in the standard β-galactosidase procedure and shown to behave in a way analogous to strain M26–26, and to be suitable as organisms to assay these particular amino acids. However, none of these was a deletion mutant, making reversion to wild type possible; in consequence, although they are perfectly acceptable when used with appropriate care, these strains are not as 'robust' as one would like for use in such an assay.

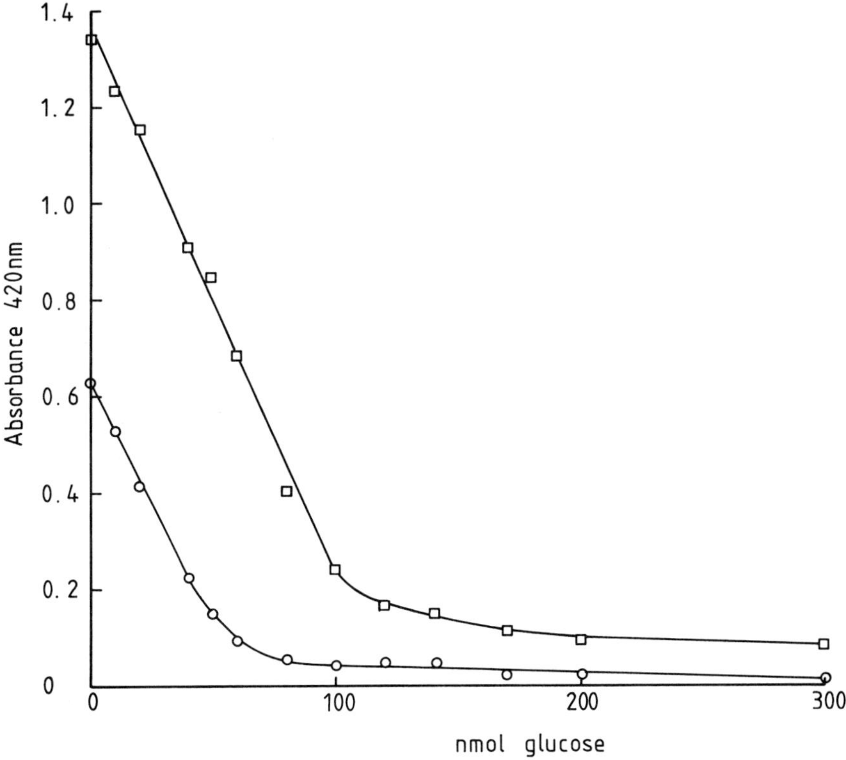

Fig. 10.3 Influence of added glucose on lysine-dependent β-galactosidase synthesis in *E. coli* K12. The quantities of glucose shown were added to two different amounts of lysine, O——O, 2 nmol lysine, □——□, 3 nmol lysine, and the resultant levels of enzyme synthesis were measured in the standard assay

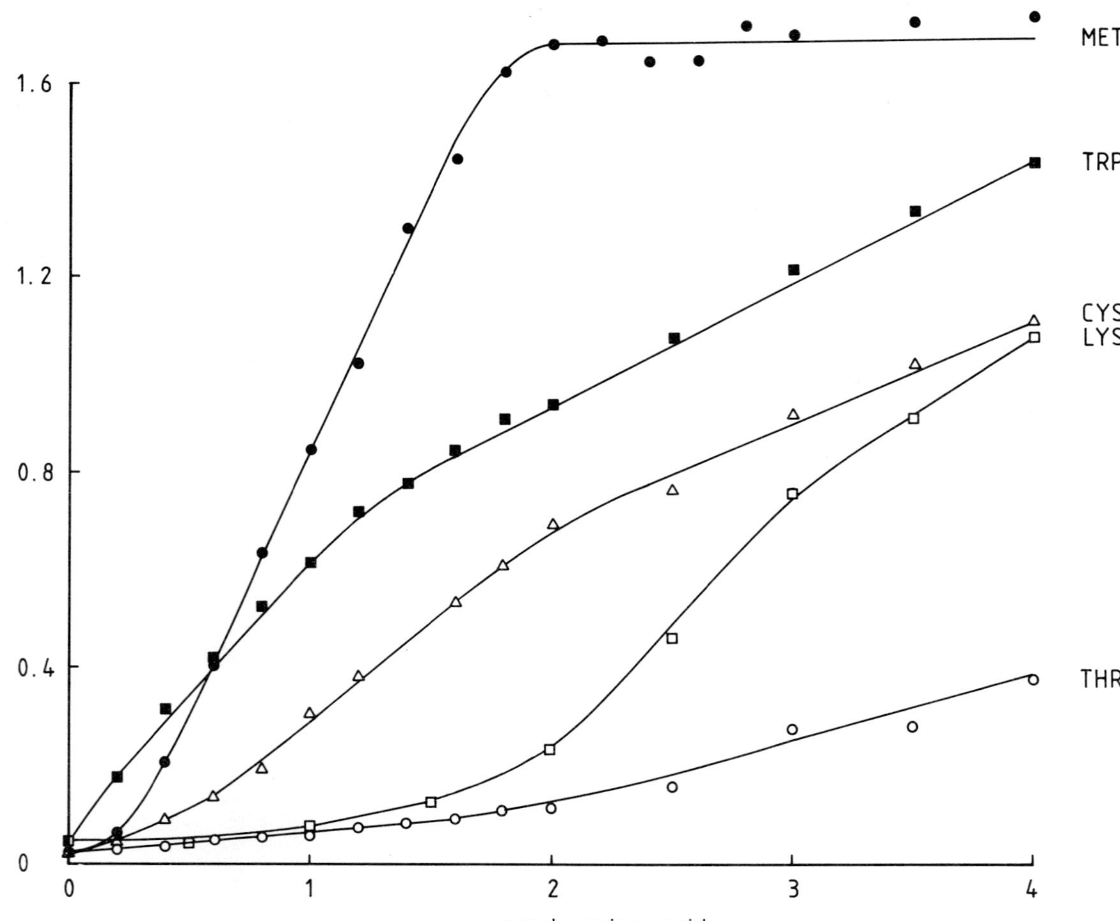

Fig. 10.4 Amino acid calibration curves using isogenic deletion mutants of *E. coli* K12. Each of the five different auxotrophs was incubated with the indicated amounts of the appropriate amino acid and β-galactosidase activity was measured as described in the text

More recently, we have constructed a collection of isogenic strains of *E. coli* K12 carrying deletions that confer auxotrophy for methionine, tryptophan, cysteine, lysine and threonine (G. M. Payne and J. W. Payne, unpublished results). The strains were produced using standard P1 transduction techniques with deletion mutants obtained from the Coli Genetic Stock Centre Yale University, New Hampshire or provided by individuals. The strains have been tested for genetic stability, turbidimetric growth response to amino acids in various forms and behaviour in the β-galactosidase assay. In Fig. 10.4 is shown the various calibration curves for all these strains. The sensitivities are broadly the same for each amino acid. The differences between the calibration curves can be attributed to various causes, e.g. different contents of the amino acids in β-galactosidase, different degrees of amino acid metabolism, etc. A full description of the construction, properties and use of these strains is in preparation.

In a preliminary report upon the use of a collection of non-isogenic deletion mutants, Hitchins *et al.* (1986) reported that all the mutants performed properly in the enzyme method. (We were kindly provided with samples of these cultures and certain were used in the construction of our own isogenic collection.) In subsequent studies (A. D. Hitchins, personal communication), these mutants have been shown to have considerable promise for the assay of available amino acids in a variety of prepared food products. However, because of their varied genetic backgrounds, occasional problems with growth and the idiosyncratic properties of certain of the mutants, on the whole these strains appear potentially less useful than our own isogenic collection.

RECENT DEVELOPMENTS AND FUTURE PROSPECTS

From the extensive studies that have been carried out by us and others using current *E. coli* assay procedures, it is clear that they offer a meaningful alternative to current animal and chemical tests. Nevertheless, there remains considerable scope for producing a more robust, convenient and commercially acceptable procedure.

Firstly, with regard to the bacterial strains, although the current collection of isogenic mutants seems acceptable for most purposes, we have now used recombinant DNA techniques to produce an additional set (G. M. Payne, I. M. Eastwood and J. W. Payne, unpublished results). Thus, isolated DNA corresponding to the structural genes of the *E. coli lac* operon has been produced in which all the associated regulatory (operator–promoter) regions have been removed by restriction endonuclease action. This DNA has then been linked to a temperature-sensitive control region derived from a mutant phage-λ element. This hybrid *lac* has been placed in a multicopy plasmid carrying an ampicillin resistance gene. The synthesis of β-galactosidase from this plasmid *lacZ* gene is not controlled by the *lac* repressor (and is therefore not inducible by IPTG) and neither is it subject to catabolite repression (and therefore is not susceptible to glucose inhibition). Transcription from the plasmid *lacZ* gene is switched off at 25°C and induced by raising the temperature to 42°C. This plasmid has been transformed into the collection of isogenic auxotrophs mentioned above. Although these still retain chromosomal *lac*, constitutive expression from this is very low in the absence of inducer, and they are convenient vehicles for investigating the temperature-controlled β-galactosidase expression; in due course a chromosomal *lac* deletion could be introduced into the test strains.

In trial studies, we have used the plasmid-containing methionine auxotroph. Bacteria are grown and then starved in glucose medium at low temperature (25°C), then simultaneously transferred to 42°C and methionine or test sample is added. Other aspects of the procedure are as usual. A typical calibration curve is obtained, which is very similar to that for the IPTG-induced synthesis. Thus, with this strain the procedure is somewhat simpler (no inducer-addition step is needed), cheaper (no IPTG is required), and free from interference from glucose (catabolite repression is absent) and by culturing in the presence of ampicillin (to maintain the plasmid) chances of bacterial contamination are minimized.

The availability of strains such as these, specifically tailored to the needs of the assay through application of recombinant DNA techniques, should ensure the bacterial component of the assay kit is optimised for general use.

Secondly, is the possibility for improvements to the actual assay procedure. It is possible that improvements could be achieved without significantly modifying the general protocol by use of alternative enzyme substrates, e.g. chlorophenol red–β-D-galactopyranoside ($\lambda_{max}=574$ nm, $\varepsilon=75\,000$) or 4-methylumbelliferyl-β-D-galactopyranoside, which allows fluorimetric measurement, but we have not tested these possibilities.

We have spent some time investigating the feasibility of producing a continuous-flow, automated system, but have now abandoned this endeavour. On the other hand, we have now successfully adapted the method for use with conventional microtitre plate readers. We have used a machine equipped with a pumping system that allows samples/reagents to be added in a programmed manner to any or all of the 96 wells of a microtitre plate. Plates may be shaken and incubated at a range of temperatures within the plate housing of the machine. Colour in the wells can be read rapidly (ca. 45 sec for all 96 wells) and repeatedly, so that kinetic end-point readings can be taken if desired, without the need to stop the enzymic reaction (by adding Na_2CO_3). Total incubation volumes are 200 μl and amount of test material needed is therefore decreased 10-fold. Sophisticated software allows the data to be stored, handled, processed and printed in a variety of forms to satisfy individual needs. We are convinced that the marrying of this technology with the appropriate protocol and strains of *E. coli* now provides a commercially attractive package of widespread applicability to the determination of available amino acids.

Summary

There is a widespread need to measure the nutritional availability of various amino acids, e.g. lysine, methionine, tryptophan, cysteine and threonine, in a range of human foods and animal feedstuffs. It is also necessary to monitor the effects of various treatments used in food-processing for their effects upon amino acid availability. Current biological (animal), chemical and microbiological procedures used in such assays are briefly reviewed and evaluated. Attention is drawn to the unnecessary limitations imposed on bacterial assays by restricting the organisms used to naturally-occurring strains. It is argued that many of these restrictions can be overcome by the use of genetically-engineered, auxotrophic strains of *Escherichia coli*, tailored to the specific demands of an assay for amino acid availability.

In testing this thesis, development of a conventional turbidimetric tube assay for lysine using an *E. coli* auxotroph is described, together with its application to the assay of a range of protein sources. Comparisons between the results obtained and those from rat and chemical assays are discussed.

In subsequent studies, using microbial genetics and recombinant DNA techniques, a collection of amino acid deletion mutants of *E. coli* has been produced. These have been used in a new colorimetric assay based on measurement of amino acid-dependent enzyme (β-galactosidase) synthesis. The procedure is absolutely specific for a given amino acid, sensitive (less than 5 nmol of amino acid is needed), rapid (about 2.5 h per assay), and results show good correlation with nutritional values obtained from rat assays. The procedure has been adapted for use with commercial microtitre plate readers.

Acknowledgements

I am indebted to G. M. Payne and J. M. Bainbridge (formerly J. M. Tuffnell) for their enthusiasm and ideas, and without whom the enzyme assay would not have been developed. The contribution of I. M. Eastwood to the recombinant DNA studies is gratefully acknowledged. A. D. Hitchins kindly provided strains and details of results prior to publication.

The financial assistance of The Agricultural and Food Research Council and, at an earlier stage, The Humane Research Trust, is gratefully acknowledged.

References

Alves, R. A., Gleaves, J. T. and Payne, J. W. (1985). The role of outer membrane proteins in peptide uptake by *Escherichia coli. FEMS Microbiol. Lett.*, **27**, 333–8.

Anantharaman, K., Gallaz, L. and Decarli, B. (1983a) Microbiological determination of available lysine in proteins using a Lys *Escherichia coli* M26–26 as an alternative to *in vivo* nutritional assay in rats. *Z. Versuchstierkunde*, **25**, 147–8.

Anantharaman, K., Parard, C., Decarli, B., Reinhardt, P., Graf, E. and Finot, P. A. (1983b). The microbiological estimation of available lysine in industrial milk products using a lysine$^-$ *Escherichia coli* M26–26. In *Research in Food Science and Nutrition*, Eds McLoughlin, J. V. and McKenna, B. M., Vol. 2, *Basic Studies in Food Science*, pp. 81–2. Dublin, Boole Press.

Bell, G., Higgins, C. F., Payne, G. M. and Payne, J. W. (1977). Use of *Escherichia coli* to determine available lysine in plant proteins. In *Nutritional Evaluation of Cereal Mutants*, pp. 107–23. Vienna, International Atomic Energy Agency.

Bigwood, E. J. (1972). Protein and Amino Acid Functions. Oxford, Pergamon Press.

Board, R. G. and Lovelock, D. W. (1975). *Some Methods for Microbiological Assay*. London, Academic Press.

Bodwell, C. E. (1976). *Evaluation of Proteins for Humans*. Westport, Connecticut, Avi Publishing Co.

Bodwell, C. E., Adkins, J. S. and Hopkins, D. T. (Eds) (1981) *Protein Quality in Humans: Assessment and* in vitro *Estimation*. Westport, Connecticut, Avi Publishing Co.

Bolinder, A. E. (1972). Large plate assays for amino acids. In *Analytical Microbiology*, Ed. Kavanagh, F., Vol. 2, pp. 479–591. New York and London, Academic Press.

Carpenter, K. J. (1960). The estimation of the available lysine in animal-protein foods. *Biochem. J.*, **77**, 604–10.

Cole, D. J. A., Boorman, K. N., Buttery, P. J., Lewis, D., Neale, R. J. and Swan, H. (Eds) (1976). *Protein Metabolism and Nutrition*. London, Butterworths.

Concon, J. M. (1975). Chemical determination of critical amino acids. In *Protein Nutritional Quality of Foods and Feeds*, Part 1, *Assay Methods*, Ed. Friedman, M., pp. 311–80. New York, Marcel Dekker.

Erbersdobler, H. F. and Anderson, T. R. (1983). Determination of available lysine by various procedures in Maillard-type products. In *The Maillard Reaction in Foods and Nutrition*, Eds Waller, G. R. and Feather, M. S., ACS Symposium Series 215. Washington, DC, American Chemical Society.

Finot, P. A., Bujard, E., Mottu, F. and Mauron, J. (1977). Availability of the true Schiff's bases of lysine. Chemical evaluation of the Schiff's base between lysine and lactose in milk. *Adv. Expl. Med. Biol.*, **86B**, 343–65.

Ford, J. E. (1981). Microbial methods for protein quality assessment. In *Protein Quality in Humans: Assessment and in vitro Estimation*, Eds Bodwell, C. E., Adkins, J. S., and Hopkins, D. T., pp. 278–305. Westport, Connecticut, Avid Publishing Co.

Friedman, M. (Ed.) (1975). *Protein Nutritional Quality of Foods and Feeds.* Parts 1 and 2. New York, Marcel Dekker.

Gruenwadel, D. W. and Whitaker, J. R. (Eds) (1985). *Food Analysis, Principles and Techniques*, Vol. 3, *Biological Techniques.* New York and Basel, Marcel Dekker.

Hewitt, W. (1977). *Microbiological Assay: An Introduction to Quantitative Principles and Evaluation.* New York and London, Academic Press.

Hitchins, A. D., McDonough, F. E. and Bodwell, C. E. (1986). Amino acid bioavailability assays by specific enzyme induction in amino acid deletion mutants of *Escherichia coli. Internat. Cong. Microbiol. Manchester*, **14**, 299, Abstracts 69–70.

Kavanagh, F. W. (1972). *Analytical Microbiology*, Vol. 2. New York, Academic Press.

Landers, R. E. and Gierhart, D. L. (1985). Use of bacteria, fungi, protozoa and yeast as analytical tools in food analysis. In *Food Analysis, Principles and Techniques*, vol. 3, *Biological Techniques*, Eds Gruenwedel, D. W. and Whitaker, J. R., pp. 35–84. New York and Basel, Marcel Dekker.

Matthews, D. M. and Payne, J. W. (1975). Peptides in the nutrition of microorganisms and peptides in relation to animal nutrition. In *Peptide Transport in Protein Nutrition*, Eds Matthews, D. M. and Payne, J. W., pp. 1–60. Amsterdam and New York, North-Holland and American Elsevier.

Matthews, D. M. and Payne, J. W. (1980). Transmembrane transport of small peptides. *Curr. Top. Memb. Transp.*, **14**, 331–425.

Payne, J. W. (1980). Transport and utilisation of peptides by bacteria. In *Microorganisms and Nitrogen Sources*, Ed. Payne, J. W., pp. 211–57. Chichester, J. Wiley.

Payne, J. W. (1986). Drug delivery systems: optimising the structure of peptide carriers for synthetic antimicrobial drugs. *Drugs exp. clin. Res.*, **12**, 585–95.

Payne, J. W. and Gilvarg, C. (1968). Size restriction on peptide utilisation in *Escherichia coli. J. Biol. Chem.*, **243**, 6291–9.

Payne, J. W. and Tuffnell, J. M. (1980). Assays for amino acids, peptides and proteins. In *Microorganisms and Nitrogen Sources*, Ed. Payne, J. W., pp. 727–65. Chichester, J. Wiley.

Payne, J. W., Bell, G. and Higgins, C. F. (1977). The use of an *Escherichia coli lys* auxotroph to assay nutritionally available lysine in biological materials. *J. Appl. Bacteriol.*, **42**, 165–77.

Porter, J. W. G. and Rolls, B. A. (Eds). (1973). *Protein in Human Nutrition.* London and New York, Academic Press.

Tuffnell, J. M. (1983). Studies on Amino Acid Assays Using *Escherichia coli.* PhD thesis, Durham University.

Tuffnell, J. M. and Payne, J. W. (1985). A colorimetric enzyme assay using *Escherichia coli* to determine nutritionally available lysine in biological materials. *J. Appl. Bacteriol.*, **58**, 333–41.

Waller, G. R. and Feather, M. S. (Eds) (1983). *The Maillard Reaction in Foods and Nutrition.* ACS Symposium Series 215. Washington, DC, American Chemical Society.

Index